I0880887

Review Criteria for Successful Treatment of Hydrolysate at the Pueblo Chemical Agent Destruction Pilot Plant

Committee on Review Criteria for Successful Treatment of Hydrolysate at the Pueblo and Blue Grass Chemical Agent Destruction Pilot Plants

Board on Army Science and Technology

Division on Engineering and Physical Sciences

NATIONAL RESEARCH COUNCIL
OF THE NATIONAL ACADEMIES

THE NATIONAL ACADEMIES PRESS
Washington, D.C.
www.nap.edu

THE NATIONAL ACADEMIES PRESS • 500 Fifth Street, NW • Washington, DC 20001

NOTICE: The project that is the subject of this report was approved by the Governing Board of the National Research Council, whose members are drawn from the councils of the National Academy of Sciences, the National Academy of Engineering, and the Institute of Medicine. The members of the committee responsible for the report were chosen for their special competences and with regard for appropriate balance.

This study was supported by Contract No. W911NF-14-1-0280 between the National Academy of Sciences and the U.S. Army. Any opinions, findings, conclusions, or recommendations expressed in this publication are those of the author(s) and do not necessarily reflect the views of the organizations or agencies that provided support for the project.

International Standard Book Number-13: 978-0-309-31788-7
International Standard Book Number-10: 0-309-31788-6

Limited copies of this report are available from the Board on Army Science and Technology, National Research Council, 500 Fifth Street, NW, Room 940, Washington, DC 20001; (202) 334-3118.

Additional copies of this report are available from the National Academies Press, 500 Fifth Street, NW, Keck 360, Washington, DC 20055; (800) 624-6242 or (202) 334-3313 (in the Washington metropolitan area); http://www.nap.edu.

Copyright 2015 by the National Academy of Sciences. All rights reserved.

Printed in the United States of America

THE NATIONAL ACADEMIES

Advisers to the Nation on Science, Engineering, and Medicine

The **National Academy of Sciences** is a private, nonprofit, self-perpetuating society of distinguished scholars engaged in scientific and engineering research, dedicated to the furtherance of science and technology and to their use for the general welfare. Upon the authority of the charter granted to it by the Congress in 1863, the Academy has a mandate that requires it to advise the federal government on scientific and technical matters. Dr. Ralph J. Cicerone is president of the National Academy of Sciences.

The **National Academy of Engineering** was established in 1964, under the charter of the National Academy of Sciences, as a parallel organization of outstanding engineers. It is autonomous in its administration and in the selection of its members, sharing with the National Academy of Sciences the responsibility for advising the federal government. The National Academy of Engineering also sponsors engineering programs aimed at meeting national needs, encourages education and research, and recognizes the superior achievements of engineers. Dr. C. D. Mote, Jr., is president of the National Academy of Engineering.

The **Institute of Medicine** was established in 1970 by the National Academy of Sciences to secure the services of eminent members of appropriate professions in the examination of policy matters pertaining to the health of the public. The Institute acts under the responsibility given to the National Academy of Sciences by its congressional charter to be an adviser to the federal government and, upon its own initiative, to identify issues of medical care, research, and education. Dr. Victor J. Dzau is president of the Institute of Medicine.

The **National Research Council** was organized by the National Academy of Sciences in 1916 to associate the broad community of science and technology with the Academy's purposes of furthering knowledge and advising the federal government. Functioning in accordance with general policies determined by the Academy, the Council has become the principal operating agency of both the National Academy of Sciences and the National Academy of Engineering in providing services to the government, the public, and the scientific and engineering communities. The Council is administered jointly by both Academies and the Institute of Medicine. Dr. Ralph J. Cicerone and Dr. C. D. Mote, Jr., are chair and vice chair, respectively, of the National Research Council.

www.national-academies.org

COMMITTEE ON REVIEW CRITERIA FOR SUCCESSFUL TREATMENT OF HYDROLYSATE AT THE PUEBLO AND BLUE GRASS CHEMICAL AGENT DESTRUCTION PILOT PLANTS

ROBERT A. BEAUDET, University of Southern California, Pasadena (retired), *Co-Chair*
TODD A. KIMMELL, Argonne National Laboratory, Washington, D.C., Office, *Co-Chair*
EDWARD J. BOUWER, Johns Hopkins University, Baltimore, Maryland
JUDITH A. BRADBURY, Independent Consultant, Knoxville, Tennessee
REBECCA A. HAFFENDEN, Argonne National Laboratory, Santa Fe, New Mexico
HANK C. JENKINS-SMITH, University of Oklahoma, Norman
KIMBERLY L. JONES, Howard University, Washington, D.C.
MURRAY G. LORD, Dow Chemical Company, Freeport, Texas
TRISHA H. MILLER, Sandia National Laboratories, Shoreview, Minnesota
ROBERT PUYEAR, Independent Consultant, Chesterfield, Missouri
WILLIAM R. RHYNE, Independent Consultant, Kingston, Tennessee
PHILLIP E. SAVAGE, Pennsylvania State University, State College
PHILIP C. SINGER, University of North Carolina at Chapel Hill (retired)

Staff

NANCY T. SCHULTE, Study Director
NIA D. JOHNSON, Senior Research Associate
DEANNA P. SPARGER, Program Administrative Coordinator

BOARD ON ARMY SCIENCE AND TECHNOLOGY

DAVID M. MADDOX, Independent Consultant, Arlington, Virginia, *Chair*
JEAN D. REED, Independent Consultant, Arlington, Virginia, *Vice Chair*
DUANE ADAMS, Independent Consultant, Arlington, Virginia
ILESANMI ADESIDA, University of Illinois at Urbana-Champaign
STEVEN W. BOUTELLE, Cisco Consulting Services, Herndon, Virginia
EDWARD C. BRADY, Strategic Perspectives, Inc., McLean, Virginia
W. PETER CHERRY, Independent Consultant, Ann Arbor, Michigan
EARL H. DOWELL, Duke University, Durham, North Carolina
JULIA D. ERDLEY, Pennsylvania State University, State College
LESTER A. FOSTER, Electronic Warfare Associates, Herndon, Virginia
JAMES A. FREEBERSYSER, BBN Technology, St. Louis Park, Minnesota
PETER N. FULLER, Cypress International, Springfield, Virginia
W. HARVEY GRAY, Independent Consultant, Oak Ridge, Tennessee
JOHN J. HAMMOND, Independent Consultant, Fairfax, Virginia
RANDALL W. HILL, JR., University of Southern California, Institute for Creative Technologies, Playa Vista
JOHN W. HUTCHINSON, Harvard University, Cambridge, Massachusetts
BRUCE D. JETTE, Synovision Solutions, Burke, Virginia
ROBIN L. KEESEE, Independent Consultant, Fairfax, Virginia
WILLIAM L. MELVIN, Georgia Tech Research Institute, Smyrna
WALTER F. MORRISON, Independent Consultant, Alexandria, Virginia
ROBIN MURPHY, Texas A&M University, College Station
SCOTT PARAZYNSKI, University of Texas Medical Branch, Galveston
RICHARD R. PAUL, Independent Consultant, Bellevue, Washington
DANIEL PODOLSKY, University of Texas Southwestern Medical Center, Dallas
LEON E. SALOMON, Independent Consultant, Gulfport, Florida
ALBERT A. SCIARRETTA, CNS Technologies, Inc., Springfield, Virginia
JONATHAN M. SMITH, University of Pennsylvania, Philadelphia
DAVID A. TIRRELL, California Institute of Technology, Pasadena
MICHAEL A. VANE, DynCorp International, Lorton, Virginia
JOSEPH YAKOVAC, JVM LLC, Hampton, Virginia

Staff

BRUCE A. BRAUN, Director
CHRIS JONES, Financial Manager
DEANNA P. SPARGER, Program Administrative Coordinator

Preface

When I had to rotate off the Board on Army Science and Technology in 1995, the program director asked me if I would like to chair the first Assembled Chemical Weapons Assessment (ACWA) Committee. The U.S. Congress had just passed Public Laws 104-201 and 104-208 establishing ACWA. Of course I said I would be happy to. From that time on, I have been involved with ACWA in one way or another.

Finally, after all these years, the Army is preparing to destroy the chemical stockpile at the Pueblo Chemical Depot. The facility, called, in full, the Pueblo Chemical Agent Destruction Pilot Plant (PCAPP), will destroy its stockpile of 155-mm and 105-mm artillery shells and 4.2-in. mortars, all of which contain one form or another of the chemical agent mustard. The munitions are robotically opened and the mustard collected. The next step in the destruction process is the neutralization of the mustard with lye to produce a product called the hydrolysate. The second step, biotreatment of the hydrolysate, is a first-of-a-kind system that has not been extensively tested with the hydrolysate. The concerns noted by the earlier committee are held by this committee as well and are summarized in this report. Thus, there is some concern that this biotreatment will not mineralize the hydrolysate to water, carbon dioxide, and salts. In that case, the Army wants to hedge its bets by considering offsite transportation and disposal of the hydrolysate. Thus, it asked the National Research Council (NRC) to form an ad hoc committee to recommend when the hydrolysate could be sent offsite. The committee has bent over backward to include and interact with the public, the stakeholders, and the Citizens' Advisory Commissions in Pueblo and Blue Grass. This report presents the committee's findings and a recommendation.

Unfortunately, during the course of this study, I developed a medical problem that prevented me from traveling and being further involved with the study. And fortunately, Todd Kimmell came to my rescue and took over the chairing of this committee. Like me, Todd has also been involved with the ACWA since the beginning of the program, but from a different perspective. He was part of the team that developed the initial environmental impact studies that supported the selection of the ACWA alternative technologies. He also has a great deal of experience with NRC committees, having been a member since 2001 of nonstockpile, stockpile, and ACWA committees. I am greatly indebted to Todd for continuing this work, and I know I leave the committee in good hands.

Todd and I and the committee thank all the PCAPP staff, including Rick Holmes, the PCAPP Project Manager; George Lecakes, the PCAPP Chief Scientist; Bruce Huenefeld, the PCAPP site manager; Paul Usinowicz, the PCAPP technical advisor; and Irene Kornelly and Ross Vincent, both members of the Colorado Citizens' Advisory Commission, for having patience with us and for answering our numerous and sometimes naïve questions.

Also, we thank the NRC staff, including the study director, Nancy Schulte; the program administrative coordinator, Deanna Sparger; and the senior research associate, Nia Johnson, for their continuous support, patience, and assistance in producing this report.

Robert A. Beaudet, *Co-Chair*
Todd A. Kimmell, *Co-Chair*

Committee on Review Criteria for Successful Treatment of Hydrolysate at the Pueblo and Blue Grass Chemical Agent Destruction Pilot Plants

Acknowledgment of Reviewers

This report has been reviewed in draft form by individuals chosen for their diverse perspectives and technical expertise, in accordance with procedures approved by the National Research Council's (NRC's) Report Review Committee. The purpose of this independent review is to provide candid and critical comments that will assist the institution in making its published report as sound as possible and to ensure that the report meets institutional standards for objectivity, evidence, and responsiveness to the study charge. The review comments and draft manuscript remain confidential to protect the integrity of the deliberative process. We wish to thank the following individuals for their review of this report:

Cheryl A. Burke, Dow Chemical Company,
Charles R. Cantor, Sequenom, Inc.,
Raymond M. Hozalski, University of Minnesota,
Douglas M. Medville, MITRE Corporation (retired),
Leonard M. Siegel, Center for Public Environmental Oversight,
Vernon L. Snoeyink, University of Illinois, and
William J. Walsh, Pepper Hamilton LLP.

Although the reviewers listed above have provided many constructive comments and suggestions, they were not asked to endorse the conclusions or recommendations nor did they see the final draft of the report before its release. The review of this report was overseen by Hyla S. Napadensky, Napadensky Energetics, Inc. Appointed by the NRC, she was responsible for making certain that an independent examination of this report was carried out in accordance with institutional procedures and that all review comments were carefully considered. Responsibility for the final content of this report rests entirely with the authoring committee and the institution.

Contents

SUMMARY 1

1 INTRODUCTION 8
Related NRC 2013 Report, 9
Need for the Present NRC Study, 9
Organization of This Report, 10
Reference, 10

2 BACKGROUND 11
Brief Description of the PCAPP Process, 11
Production and Characterization of Hydrolysate, 12
Description of the Immobilized Cell Bioreactors, 14
Description of the Brine Reduction System, 14
References, 16

3 STAKEHOLDER INTERESTS AND ISSUES 17
Background, 17
Methodology, 17
Summary, 19
Findings and Recommendations, 19
References, 20

4 REGULATORY REQUIREMENTS FOR OFFSITE HYDROLYSATE SHIPMENT AND TREATMENT 21
Introduction, 21
RCRA Permitting, 21
NEPA Requirements, 25
Pueblo County Certificate of Designation, 26
Organisation for the Prohibition of Chemical Weapons, 27
References, 28

5 TRANSPORTATION OF CHEMICAL MATERIALS 29
Regulations of the U.S. Department of Transportation, 29
Historical Transportation of Chemical Munition Materials, 30
Identification of Hydrolysate Transportation Risks, 30
Related Prior NRC Findings and Recommendations, 34
References, 34

6 HYDROLYSATE TREATMENT PERFORMANCE GOALS 35
Decision Framework for Determining Successful Plant Operation, 35
Performance Criteria for Hydrolysate, 36
Preoperational Testing Data That Factor into the Decision Process, 39
References, 40

7 FAILURE RISKS, SYSTEMIZATION, AND CONTINGENCY OPTIONS 41
Failure Risks, Systemization, and Contingency Options in the Immobilized Cell Bioreactor, 41
Failure Risks, Systemization, and Contingency Options in the Water Recovery and Brine Reduction Systems, 45
Offsite Shipment as a Contingency Option, 49
References, 51

APPENDIXES

A Statement of Task 55
B Public Interest and Input Documents 56
C Biographical Sketches of Committee Members 67
D Committee Meetings 70

Tables, Figures, and Box

TABLES

S-1 Graded Success Scale for Use in Evaluating Overall Operation and Individual Treatment Processes (ICBs, WRS, BRS), 3
S-2 Summary of Potential Technical Factors Leading to System Failure in the Immobilized Cell Bioreactor Units, and Contingency Options, 5
S-3 Summary of Potential Technical Factors Leading to System Failure in the Brine Reduction System, and Contingency Options, 6

1-1 Inventory of Assembled Chemical Weapons at Pueblo Chemical Depot, 9

2-1 HD Hydrolysate Characterization from 2003 Biotreatment Testing and RCRA RD&D Permit Waste Analysis Plan, 13
2-2 Key Design Operating and Feed Characteristics for the Immobilized Cell Bioreactor Units, 14
2-3 Issues Identified in a Review of the Immobilized Cell Bioreactors and Biotreatment of the Hydrolysate, 15
2-4 Issues Identified in a Review of the Brine Reduction System, 15

5-1 Comparison of Chemical Agent Liquid Treatment Content, 31
5-2 Historical Shipment Data, 32
5-3 Estimated PCAPP Truck Shipments, 33
5-4 Highway Hazmat Incident Summary by Transportation Phase in 2013, 33

6-1 Graded Success Scale for Use in Evaluating Overall Operation and Individual Treatment Processes (ICBs, WRS, and BRS), 39
6-2 Description of How Data Generated During Systemization Could Be Used to Reduce the Risk of Ineffective Operation or System Failure, and Possible Alternatives to Reduce These Risks, 40

7-1 Summary of Potential Technical Factors Leading to System Failure in the Immobilized Cell Bioreactor Units, and Contingency Options, 46
7-2 Summary of Potential Technical Factors Leading to System Failure in the Brine Reduction System, and Contingency Options, 49

FIGURES

2-1 PCAPP biotreatment area process diagram, 12

4-1 Transitioning of RCRA permit, 23

6-1 Decision framework, 37

7-1 PCAPP BTA process diagram, 42

BOX

6-1 Performance Criteria for Hydrolysate, 38

Acronyms and Abbreviations

ABCDF	Aberdeen Chemical Agent Disposal Facility
ACWA	Assembled Chemical Weapons Alternatives[1]
APG	Aberdeen Proving Ground
BGAD	Blue Grass Army Depot
BGCAPP	Blue Grass Chemical Agent Destruction Pilot Plant
BGCDF	Blue Grass Chemical Disposal Facility
BRS	brine reduction system
BTA	biotreatment area
CAC	Citizens' Advisory Commission
CDPHE	Colorado Department of Public Health and Environment
COD	chemical oxygen demand
CSEPP	Chemical Stockpile Emergency Preparedness Program
CSTR	continuous-flow stirred tank reactor
CWC	Chemical Weapons Convention
DAP	diammonium phosphate
DOT	Department of Transportation
EIS	environmental impact statement
EPA	Environmental Protection Agency
FAA	Federal Aviation Administration
FEA	Final Environmental Assessment
FEIS	Final Programmatic Environmental Impact Statement
FEMA	Federal Emergency Management Program
FMCSA	Federal Motor Carrier Safety Administration
FRA	Federal Railroad Administration
GAC	granular activated carbon
GB	nerve agent (sarin)
H	Levinstein mustard agent
HD	distilled mustard agent
HT	distilled mustard mixed with bis(2-chloroethylthioethyl) ether
ICB	immobilized cell bioreactor
LDR	Land Disposal Restriction (RCRA)
MCL	Maximum Contaminant Level
MINICAMS	miniature continuous air monitoring system(s)
MWS	munition washout station
NECD	Newport (Indiana) Chemical Depot
NECDF	Newport (Indiana) Chemical Agent Disposal Facility
NEPA	National Environmental Policy Act
NRC	National Research Council
OPCW	Organisation for the Prohibition of Chemical Weapons
PCAPP	Pueblo Chemical Agent Destruction Pilot Plant
PCD	Pueblo Chemical Depot
PEO ACWA	Program Executive Office for Assembled Chemical Weapons Alternatives
PHMSA	Pipeline and Hazardous Materials Safety Administration
QTRA	quantitative transportation risk analysis
RCRA	Resource Conservation and Recovery Act
RD&D	Research and Development and Demonstration (RCRA)
REC	record of environmental consideration

[1] Before June 2003, Assembled Chemical Weapons Assessment.

RMA	Rocky Mountain Arsenal
TDG	thiodiglycol
TDS	total dissolved solids
TOC	total organic carbon
TSDF	treatment, storage, and disposal facility
TSS	total suspended solids
VSS	volatile suspended solids
VX	nerve agent
WRS	water recovery system

Summary

One of the last two sites with chemical munitions and chemical materiel is the Pueblo Chemical Depot (PCD) in Pueblo, Colorado. The stockpile at PCD consists of about 800,000 projectiles and mortars, all of which are filled with the chemical agent mustard. Under the direction of the Assembled Chemical Weapons Alternatives (ACWA) program, the Army has constructed the Pueblo Chemical Agent Destruction Pilot Plant (PCAPP)[1] to destroy these munitions. The primary technology to be used to destroy the mustard agent at PCAPP is hydrolysis, resulting in a secondary waste stream referred to as *hydrolysate*.

PCAPP features a process that will be used to treat the hydrolysate and the thiodiglycol (TDG)—a breakdown product of mustard—contained within. The process is a biotreatment technology that uses what are known as immobilized cell bioreactors (ICBs). After biodegradation, the effluent flows to a brine reduction system (BRS), producing a solidified filter cake that is intended to be sent offsite to a permitted hazardous waste disposal facility. Water recovered from the brine reduction system is intended to be recycled back through the plant, thereby reducing the amount of water that is withdrawn from groundwater. These processes will occur within the biotreatment area (BTA) of PCAPP. The entire process is detailed in Chapter 2.

While hydrolysis itself is a proven technology, as is biotreatment, never before have these technologies been combined. Considering the first-of-a-kind nature of the application of this combination of technologies for destruction of the mustard at PCAPP and TDG within the hydrolysate, ACWA program officials have been concerned that the operation may not function as designed, and have been particularly concerned with the back end of the process, biotreatment followed by brine reduction and water recovery. ACWA commissioned a National Research Council (NRC) study, completed in 2013, *Review of Biotreatment, Water Recovery, and Brine Reduction Systems for the Pueblo Chemical Agent Destruction Pilot Plant*. The authoring committee identified a number of concerns in this report, but, overall, it had no overarching concerns that the process would not work on mustard hydrolysate.

The ACWA program managers and the PCAPP facility, including its contractor design, construction, and operations staffs, believe that the facility will perform successfully. The NRC committee writing this report believes that there is a high probability that the PCAPP facility should be able to perform successfully. However, there is still a possibility that the biotreatment, water recovery, and/or brine reduction processes may not perform satisfactorily.

In the event that one or more of these systems is shut down, even for a short period of time, destruction of the primary stockpile at PCD may need to be halted unless there is sufficient storage capacity for hydrolysate while agent hydrolysis continues, or there is an alternative means for treatment of the hydrolysate. The committee believes that destruction of the stockpile at PCD must continue, because it is destruction of the munitions and the agent that will reduce the primary risk to the local community. Hence, even though the PCAPP facility is expected to be operated successfully, it is prudent, even necessary, to establish a backup plan. Installing additional hydrolysate storage capacity is an option but would require additional regulatory permitting, and there may be a limit to how much or how long hydrolysate can be stored.

Finding 1-3. Destruction of the munitions and the agent will eliminate the primary risk to the local community. Hence, even though the PCAPP facility is expected to perform successfully, it will be prudent, even necessary, to establish a backup plan—an alternative to the onsite treatment processes intended for the hydrolysate.

[1] PCAPP is named a pilot plant because some of the processes used for destroying the agent and munition bodies have not been used, or used in combination with each other, before.

An alternative to onsite hydrolysate treatment that may be quickly implementable would be to ship the hydrolysate offsite to an existing prequalified, permitted treatment, storage, and disposal facility (TSDF). To study this alternative, the ACWA program asked the NRC to form an ad hoc committee, the Committee on Review Criteria for Successful Treatment of Hydrolysate at the Pueblo and Blue Grass Chemical Agent Destruction Pilot Plants (referred to in this report as "the committee"), to assess the PCAPP process and the potential for offsite transport of the hydrolysate. The committee's statement of task can be found in Appendix A.

STAKEHOLDER CONCERNS

Key in the offsite decision process is to consider the concerns of the local community. As explained in Chapter 3, the local citizenry in the Pueblo area are represented by a Citizens' Advisory Commission (CAC), formed in 2003, that is composed of nine members appointed by the state governor. The CAC has been and continues to be the focal point for public discussion of PCAPP issues.

The committee held its first meeting with ACWA and PCAPP representatives in Pueblo in July 2014 to facilitate local attendance. It had purposely scheduled the meeting during a period when the CAC had scheduled one of its public meetings. In this manner, NRC committee members were able to attend the CAC meeting, introduce committee members, provide a brief overview of the study, respond to questions, and emphasize the importance of community input. Equally important, CAC members were invited to join the meetings with ACWA and PCAPP, where they could listen in on the presentations and participate in open discussion. Two members of the CAC, including the chair, attended the 2-day open meetings.

From these interactions, it has become clear to the committee that the CAC and PCAPP staff have developed a sound working relationship. The committee believes that this working relationship will serve as a strong foundation for a credible consultation process should issues arise with operation of the PCAPP and the BTA.

The committee also learned during these interactions that the CAC continues to maintain its long-standing opposition to offsite shipment of hydrolysate. The committee recognizes the concern of the CAC, especially members' skepticism concerning the need for offsite shipment of hydrolysate, but it also believes that the PCAPP facility and the ACWA community are firmly behind the commitment to make the hydrolysis, biotreatment, and brine reduction processes work. Nevertheless, the committee also believes that a backup plan is needed.

At the same time, it is prudent and even necessary for PCAPP officials and ACWA to maintain discussions with the CAC and put in place an institutional mechanism that would aim to ensure regular, open communication throughout operations and help to avoid the potential for misinterpretation of motives and decisions, thereby enhancing the probability of achieving common program goals, despite different priorities. Such a mechanism would be supplementary to, yet part of, the CAC and would build on the sound relationships that PCAPP and the CAC have worked so hard to develop.

Recommendation 3-1. In consultation with the CAC, ACWA should institutionalize an explicit consultation process that focuses on the potential for offsite shipment. This process should be established immediately and give stakeholders a clearly defined and meaningful role. The consultation process should (1) be supplementary to the more general role of the CAC; (2) provide to the CAC regular updates on the status of operations as they bear on the possible need for offsite shipments; and (3) be explicitly designed to ensure there are no surprises on the part of stakeholders if they are called on to consider offsite shipments.

REGULATORY ISSUES

As discussed in Chapter 4, regulatory requirements for offsite hydrolysate shipment and treatment are complex. Requirements stem from the Resource Conservation and Recovery Act (RCRA), as administered by the Colorado Department of Public Health and Environment (CDPHE), the National Environmental Policy Act (NEPA), and the Pueblo County Board of County Commissioners Certificate of Designation. In addition, the Organisation for the Prohibition of Chemical Weapons (OPCW) has requirements applicable to the treatment of the hydrolysate, whether onsite or offsite.

Should the offsite shipment of hydrolysate become necessary, it is clear that PCAPP's RCRA permit would need to be modified. Approval of a permit modification might take 3 to 6 months or more. Further, additional NEPA documentation might be needed to support the offsite option, as this option was never fully evaluated in the PCD/PCAPP environmental impact analyses performed earlier in support of the PCAPP technologies. The NEPA documentation process, if necessary, is also time-consuming and may take months, depending on the level of controversy.

Recommendation 4-2. PCAPP should process a permit modification for the RCRA Research and Development and Demonstration (RD&D) permit that would allow for the offsite transport of hydrolysate as a backup plan. The modification application should contemplate a temporary authorization for site preparation, preconstruction, and similar activities while PCAPP is operating under the RD&D permit.

RCRA permit modifications and NEPA documentation that support the backup plan of offsite shipment must be in place as soon as practical, and all regulatory requirements must be identified and prebriefed with the CDPHE, the CAC, ACWA, and Pueblo County, so that should the decision be made that there is no other option, implementation can be rapid, with no delay for the destruction mission.

TRANSPORTATION ASSESSMENT

One of the primary concerns regarding potential offsite transportation of hydrolysate is the risk of a transportation accident. Chapter 5 summarizes previous offsite shipments of hydrolysate from other chemical demilitarization facilities (Aberdeen Proving Ground, Maryland, and Newport Chemical Depot, Indiana), and offsite shipments of similar agent-associated fluids such as waste liquids from operation of the explosive destruction system. This summary demonstrates that hydrolysate and similar fluids have been shipped offsite without incident many times in the past.

That is not to say that a transportation accident could not occur, however. But it is important to understand that the hazard here is not from the presence of mustard within the hydrolysate; during the hydrolysis process, mustard is reduced to levels that are below the detection limit of sophisticated analytical instruments. And, although the presence of TDG is a concern, the primary hazard during transportation of hydrolysate comes from the caustic nature of the hydrolysate.

PCAPP hydrolysate may or may not be considered a Class 8 (corrosive) material, but for purposes of risk identification, Class 8 is assumed.[2] Examples of Class 8 materials are hydrochloric acid, nitric acid, sulfuric acid at a concentration of >51 percent, and solid sodium hydroxide.

The hazards due to hydrolysate exposure are modest compared to exposure to materials such as concentrated sodium hydroxide, a typical Class 8 material, which may be considered a greater hazard. While the hydrolysate risks may be considered moderate, the committee concurs with a previous committee, NRC (2008), which recommended that ACWA perform a quantitative transportation risk assessment for hydrolysate, including a quantitative assessment of the human health consequences of hydrolysate. It should also prepare a prototypical emergency response plan for hydrolysate shipment (NRC, 2008). These documents will help facilitate discussions with the public and regulators about the possible alternative of shipping hydrolysate offsite. As with regulatory documentation, the transportation assessment and emergency response plan should be prepared as a backup plan and be ready to go should it be determined that offsite transport of hydrolysate is needed (NRC, 2008).

CRITERIA FOR SUCCESS FOR PCAPP

Success for PCAPP operations is defined in Chapter 6. The primary criteria for successful treatment of hydrolysate are meeting RCRA permit requirements and ACWA requirements for treatment of the hydrolysate. This includes the production of a filter cake that meets regulatory requirements that the cake contain no free liquids and the production of process water from the BRS that is of good enough quality that it can be recycled to the plant. In addition, the overall schedule for destruction of the munitions at PCAPP must be met.

Chapter 6 sets up a decision process for evaluating treatment alternatives. It includes a number of criteria for evaluating these alternatives, including the offsite option for hydrolysate treatment. A graded evaluation of system risk allows stakeholders to qualitatively rate the potential for overall program success at any point in the project. This type of graded evaluation will facilitate communication between stakeholders and allow them to track and document PCAPP progress in a transparent and consistent way throughout the course of the project. Table S-1 exemplifies a graded scale for success that could be used for the PCAPP project.

Finding 6-4. In its recent white paper on risk reduction and mitigation, PCAPP has done a thorough job of identifying potential failure risks and providing targeted strategies to mitigate these risks in the BTA (PCAPP, 2014). Employing the decision-making framework outlined previously, the overall systemization plan, and the BTA risk reduction

TABLE S-1 Graded Success Scale for Use in Evaluating Overall Operation and Individual Treatment Processes (ICBs, WRS, and BRS)

Grade	Definition
0	Success is practically certain (very low possibility of project failure): Operations are proceeding as expected. No PCAPP actions needed.
1	High likelihood of success (low possibility of project failure): Actions should be taken by PCAPP to prepare ahead of time for implementation of contingencies in the event of failures. For example, PCAPP might begin to prepare permit modifications and planning documents, including building plans for piping and shipping.
2	Success is uncertain (moderate possibility of project failure): Actions should be taken to prepare for implementation of contingency operations. For example, PCAPP might begin processing environmental documentation and finalizing contingency plans, purchasing needed materials, and implementing changes to the infrastructure.
3	Success is unlikely with current operations (high possibility of failure of the project): Actions are taken to accelerate the implementation of contingency operations. For example, construction of needed facilities is completed as quickly as possible, and environmental approvals are expedited if they have not already been obtained.

NOTE: WRS, water recovery system.

[2] Class 8 hazmat is defined in 49 CFR 173.136 as a liquid or solid that causes (1) full thickness destruction of human skin within a specified period of time or (2) a specified corrosion rate of steel or aluminum.

and mitigation plan provides targeted strategies for PCAPP to mitigate any operational problems that become apparent during surrogate testing and systemization.

FAILURE RISKS AND CONTINGENCY OPTIONS

In Chapter 7, the committee begins with the graded scale for success that it introduced in Chapter 6 and discusses a number of potential failure risks within the BTA at the PCAPP facility. PCAPP itself identified several such sources of possible failure in the BTA along with contingency options in its aforementioned white paper on risk reduction and mitigation. PCAPP's plan is to troubleshoot the majority of these risks during facility systemization.

The hydrolysate generated by agent neutralization is a unique and complex mixture. While ICBs have been used successfully in the past to treat complex hazardous organic wastes, they have not been used to treat mustard hydrolysate, aside from bench-scale testing by ACWA. Hence, there are a number of technical factors that could lead to incomplete hydrolysate treatment. These include hydrolysate toxicity to microbial biomass, the need for careful pH and temperature control, nutrient and oxygen limitations, biomass buildup and sloughing, start-up and acclimation issues, and release of odorous compounds. The committee believes that to address these factors, which could inhibit efficient ICB operations, PCAPP should develop risk mitigation plans. These plans need to be in place prior to system start-up so that agent neutralization operations are not delayed.

Another potential issue with the BTA is the complexity of the hydrolysate feed. Biodegradation of the hydrolysate has been carried out in a laboratory setting, but it has never been done with ICBs under full-scale operating conditions. To address this, PCAPP plans to test a number of key process variables, identify potential failure points, and determine optimal ways to operate the downstream processes at PCAPP. The culmination of these measures is testing with actual TDG in a surrogate hydrolysate. This testing will be conducted under full-scale conditions, which the committee believes should allow for rapid start-up and steady-state operation of the ICBs when munitions processing begins. Nevertheless, PCAPP should have contingency plans in place prior to start-up, ready to be implemented immediately should ICB operation be suboptimal during the risk reduction and mitigation testing with the surrogate hydrolysate.

Recommendation 7-1. PCAPP should develop contingency plans to mitigate risk in the event that one or more of the above factors inhibits efficient ICB operations. Such plans should be in place prior to system start-up so that agent neutralization operations are not unduly delayed.

The committee believes that some operational strategies could be implemented in the unlikely event of insufficient biotreatment or if operational problems arise. The technical factors leading to insufficient treatment in the ICBs along with the contingency options are summarized in Table S-2. Each factor is also evaluated against the performance criteria described in Chapter 6 and assigned to a performance category based on the overall risk to PCAPP operations.

Similar to the ICBs, there are also failure risks with operation of the WRS and the BRS, and there are also contingency options that may be taken to address these risks. As explained in Chapter 7, downstream from the ICBs, the WRS and BRS will enable PCAPP to recover and recycle most of the process water into munitions processing. The WRS primarily serves as a holding tank where effluent from the ICBs and other processes are collected, mixed, and stored before being transferred to the BRS. Aside from addition of acid and stripping of carbon dioxide, no treatment or processing takes place in the WRS. As a result, failure risks and contingency options are identified only for the BRS.

Technical factors that may lead to insufficient treatment of the ICB effluent include liquid droplet carryover in the evaporator and crystallizer; failure or excessive replacement frequency of the granular activated carbon (GAC) adsorbers; high chloride content leading to corrosion; excessive biomass or organic compounds leading to fouling, foaming, or odors; and excess liquid in the filter cake. PCAPP should develop risk mitigation plans in the event that one or more of the above factors inhibits efficient BRS operations. As with the ICBs, these plans need to be in place prior to system start-up.

Recommendation 7-3. PCAPP should develop contingency plans to mitigate risk in the event that one or more of the above factors inhibit efficient BRS operations. Such plans should be in place prior to system start-up so that agent neutralization operations are not unduly delayed.

If the BRS does not perform as designed, recycling the water within the plant at PCAPP may be problematic. This failure will place much greater strain on the aquifer from which PCAPP withdraws water. Moreover, the effluent from the ICBs will also need to be shipped offsite for treatment and disposal. Because the hydrolysate is diluted eightfold prior to entering the ICBs, the liquid volume leaving the ICBs is much larger than the original hydrolysate volume. Therefore, in the event of BRS failure, the committee believes that it would be prudent to consider shipping undiluted hydrolysate offsite for treatment and disposal rather than continuing to operate the ICBs on-site. This action would minimize the total volume of material that needs to be shipped offsite and it would minimize the fresh water intake by the plant.

The BRS is expected to operate as planned, but there may be some issues that, while serious, could be mitigated and would not result in total BRS failure. The technical factors leading to incomplete treatment in the ICBs, their impacts, and contingency options are summarized in Table S-3. Each factor is also evaluated against the performance criteria

TABLE S-2 Summary of Potential Technical Factors Leading to System Failure in the Immobilized Cell Bioreactor Units, and Contingency Options

Technical Factor	Grade	Rationale for Assigned Grade	Contingency Option
TDG toxicity	0 to 1	TDG will be diluted; respirometry will identify toxicity limits. Systemization with TDG will verify treatability.	Reduce TDG loading and/or reduce flow rate to the ICBs.
Inability to control pH	1 to 2	Hydrolysate pH will be neutralized with H_2SO_4; acid generated within ICBs will be neutralized with NaOH. Systemization with TDG will verify pH control capability.	Buffer with sodium carbonate as an alternative.
Inability to control temperature	0 to 1	Steam injection ports exist for heating reactors during cold weather. No contingency as yet for cooling during hot weather, if needed.	Reduce hydrolysate throughput to accommodate slower kinetics of biodegradation during high summer temperatures.
Start-up difficulty/acclimation	1 to 2	Systemization with TDG should facilitate smooth transition to hydrolysate treatment.	Halt start-up to address problems with hydrolysate feed.
Nutrient limitations	0 to 1	Urea will be added as a source of N; DAP will be added as a source of P. Precipitation of $FePO_4$ may limit P availability; systemization with TDG will verify P availability.	DAP may be added directly to the process water recirculation line, or higher amounts of DAP may be added to the feed tank.
Oxygen limitations	0 to 1	Air will be supplied by coarse bubble diffusers to all ICB chambers to meet oxygen demand of TDG biodegradation. Systemization with TDG will verify need to redistribute influent TDG load or oxygen supply.	If oxygen demand of TDG in first chamber exceeds oxygen supply, the TDG influent can be distributed uniformly across all ICB chambers. The urea and DAP as sources of nutrients both exert an oxygen demand. Switching to nitrate and phosphate salts will eliminate this oxygen demand.
Loss/sloughing of biomass solids	1 to 2	Biomass is immobilized in ICBs, so continuous loss of biosolids should be limited; systemization with TDG surrogate will verify biomass retention and potential losses.	Increase retention time in ICBs to ensure sufficient TDG biodegradation.
Buildup of biomass solids	1 to 2	Biomass sloughing should occur naturally; systemization with TDG will help verify that solids do not build up to undesirable levels.	Increase retention time in ICBs to reduce the TDG loading rate, which will reduce the amount of biomass accumulation and sloughing.
Limited hydrolysate storage capacity	2 to 3	30-day capacity available to store hydrolysate from agent neutralization. If kinetics of biotreatment are inhibited, rate of agent neutralization can be slowed. This is not expected to be a regular occurrence but may happen intermittently.	Reduce rate of agent neutralization as needed. Construct more hydraulic buffer (storage) capacity.
Release of malodorous compounds	1 to 2	GAC adsorbers are in place to remove volatile compounds.	Install additional GAC capacity.

NOTE: DAP, diammonium phosphate.

described in Chapter 6 and assigned to a performance category based on the overall risk to PCAPP operations.

The committee believes that PCAPP has adequately researched potential issues with BTA operations and believes that the risk reduction and mitigation measures to be conducted during systemization will help identify these issues. And while it believes that, overall, PCAPP is well positioned for successful operations, the contingency measures identified above would help to resolve any issues quickly.

OFFSITE SHIPMENT AS A CONTINGENCY OPTION

The committee acknowledges that there are many uncertainties surrounding the start-up and performance of each separate component within the BTA and that one or more contingency options may have to be implemented. Each decision may have to consider a continuum of options, from quick operational tweaks to improve performance (e.g., changing chemicals to maintain pH levels), to more long-term operational changes (e.g., longer retention times) and

TABLE S-3 Summary of Potential Technical Factors Leading to System Failure in the Brine Reduction System, and Contingency Options

Technical Factor	Grade	Rationale for Assigned Grade	Contingency Options
BRS product water quality is acceptable for reuse but does not meet permit requirements.	2 to 3	Product water may not consistently meet RCRA permit requirements.	Initiate obtaining permit modification to adjust the BRS water treatment requirements.
Liquid droplet carryover through the entrainment separators.	1 to 2	Liquid droplets in the evaporator could damage compressor. In the crystallizer, they could reduce recycle water quality.	Upgrade de-entrainment devices.
Poor performance of GAC adsorbers.	0 to 1	High TOC, suspended solids, or microbial growth on GAC lead to need for backwash due to the large pressure drop across adsorbers or frequent replacement of GAC.	Install additional GAC capacity, standby GAC adsorbers.
Corrosion, especially on heat transfer surfaces.	1 to 2	High chloride content and high temperatures could lead to corrosion in crevices and under deposits.	Aggressive corrosion monitoring and deposit removal are required to avoid unexpected failures.
Foaming and fouling, especially on heat exchangers.	1 to 2	Biomass carryover from ICB can lead to acid hydrolysis of biomass and solubilization of organic matter; currently unknown how much biomass will be carried over.	Install clarifier after ICBs to remove biomass and other TSS; monitor antiscalant effectiveness and fouling tendencies; add antifoaming agents; increase cleaning frequency.
Filter cake with biological activity and organic material.	0 to 1	Biological activity may lead to unacceptable odors.	Include additives (e.g., fly ash or lime) in filter cake to inhibit biological activity.
Liquid content of filter cake is too high.	0 tol	Treatment in BRS does not yield a solid product owing to high organic content; drying agents used as additives are insufficient to dry the cake.	Find an alternate TSDF that will accept waste.
pH control.	0 to 1	Low pH is required for CO_2 stripping in the WRS, and a neutral pH is required for the BRS to minimize the corrosion potential.	Control pH in the WRS with sulfuric acid and control pH in the BRS with NaOH.

NOTE: TSDF, treatment, storage, and disposal facility; TOC, total organic content; TSS, total suspended solids.

infrastructure changes (e.g., installing a clarifier) to accommodate performance issues, to interim actions while other contingency options are being implemented (e.g., constructing and employing additional hydrolysate storage capacity), to, finally, instituting offsite shipment of hydrolysate.

The committee believes that the optimum outcome is that the existing BTA operates without the need to implement the offsite option. It considers offsite shipment of hydrolysate to be the last resort, the final option on the continuum. However, if offsite shipment of hydrolysate is implemented, one very crucial decision that will need to be made is whether the offsite shipment is temporary or permanent. The committee acknowledges the possibility that once the decision to implement offsite hydrolysate shipment is made, it may be necessary to make that process permanent due to cost, the need for stability, or other considerations. The committee also acknowledges that the fix or set of fixes needed for the BTA might take only a few days, or weeks, or even a month or two, and that it might be possible, after some delay, to start the process again and continue with onsite hydrolysate treatment.

Implementing offsite transport of hydrolysate will affect plant, paper, and people, as discussed in Chapter 7, and the effort to implement offsite transport will be considerable. If offsite transport is implemented as a temporary fix, with the intent of restarting the BTA processes, the effort to switch back to the BTA would also be considerable. Depending on the length of the delay and whether staff furloughs or layoffs have occurred, original staff may no longer be available. Besides, if the BTA processes are restarted, there is no guarantee that the fix will even work, and PCAPP may need to restart offsite shipment again. Still, the committee believes that there may be circumstances under which restarting the BTA processes, after some delay, may be feasible. The committee discussed at length whether a change to offsite shipment could be temporary, or whether this change should be permanent. However, the committee acknowledges that at this time it is impossible to predict the exact circumstances of a failure once the plant enters systemization or actual operations. It therefore concluded that it would make no specific recommendation concerning the exact nature, extent, or permanence of any option, including implementation of permanent, offsite shipment of hydrolysate.

Recommendation 7-7. To preserve the ability to ship hydrolysate offsite for treatment in the event that offsite shipment is found to be the only viable option, steps should be taken as soon as possible. Examples of such steps include initiating permit modifications; drafting alternative standard operating procedures; preparing transportation risk documentation; designing process safety controls, spill containment, and fall protection for hydrolysate loading facilities; and communicating with stakeholders about if and when this option would be utilized, including how the stakeholders would be involved in the decision process.

REFERENCES

NRC (National Research Council). 2008. *Review of Secondary Waste Disposal Planning for the Blue Grass and Pueblo Chemical Agent Destruction Plants.* Washington, D.C.: The National Academies Press.

PCAPP (Pueblo Chemical Agent Destruction Pilot Plant). 2014. *White Paper Bio-Treatment Area Risk Reduction and Mitigation.* 24852-30H-BTA-V0001. Rev. 000. April.

1

Introduction

The U.S. effort to destroy its chemical weapons and materiel was already well under way when, in 1993, it signed the Chemical Weapons Convention (CWC),[1] an international treaty outlawing the production, stockpiling, and use of chemical weapons. The weapons and chemical materiel at five of the nine U.S. storage sites have now been destroyed by robotically opening the munitions, then removing, collecting, and incinerating the chemical agent, and at two other storage sites by hydrolyzing the agent with hot water or caustic. The remaining two sites with chemical munitions and chemical materiel are the Pueblo Chemical Depot (PCD) in Pueblo, Colorado, and the Blue Grass Army Depot (BGAD) in Richmond, Kentucky.

In 1996, in response to local opposition to the use of incineration, the U.S. Congress passed Public Laws 104-201 and 104-208. These laws froze funds for construction of chemical agent destruction facilities at PCD and BGAD and directed the Army to demonstrate at least two alternatives to incineration for the destruction of the agent. Thus, in 1996, a program then called the Assembled Chemical Weapons Assessment (ACWA) program was established to evaluate other means of destroying the chemical agent.

The ACWA program manager asserted early on that stakeholders would have their voice considered in the decision-making process. During the initial phases of the ACWA program, a panel called the Dialogue Group was established to give stakeholders a voice in all decision making. The Dialogue Group included representatives of local citizens, federal, state, and local regulators, the Army, and the National Research Council (NRC). After the technologies had been selected, the Dialogue Group was disbanded in favor of Citizens' Advisory Commissions (CACs), which were based in Colorado and Kentucky. Both CACs include former members of the Dialogue Group. The ACWA program has resulted in the selection of alternatives to incineration at the two sites and has since June 2003 been referred to by the same acronym, ACWA, but with a slightly different wording: the Assembled Chemical Weapons Alternatives program.

The stockpile at PCD consists of about 800,000 projectiles and mortars, all of which are filled with the chemical agent mustard. The munitions consist of 105-mm and 155-mm artillery shells and 4.2-in. mortars. The total amount of chemical agent is approximately 2,600 tons. Two forms of mustard are included: HD, distilled mustard, with the chemical formula $Cl\text{-}CH_2\text{-}CH_2\text{-}S\text{-}CH_2\text{-}CH_2\text{-}Cl$, and HT, an ether form of HD, $(Cl\text{-}CH_2\text{-}CH_2\text{-}S\text{-}CH_2\text{-}CH_2)_2O$. All the projectiles and three-quarters of the mortars contain HD; the rest of the mortars contain HT. All the munitions and their quantities are listed in Table 1-1.

The facility to destroy the munitions at PCD is called the Pueblo Chemical Agent Destruction Pilot Plant (PCAPP). At the writing of this report, the plant has been constructed and is now completing systemization. Risk reduction and mitigation studies are being conducted concurrent with systemization. Processing of the mustard munitions through the plant is scheduled to begin in September 2015.[2] It is expected that it will require between 4 and 5 years to completely destroy the PCD stockpile. The PCAPP process involves hydrolysis of the mustard, followed by biotreatment of the residual, known as hydrolysate, in immobilized cell bioreactors (ICBs), and treatment of the ICB effluent in a brine reduction system (BRS). While destruction of the mustard itself is conducted under the auspices of the CWC, because the hydrolysate contains thiodiglycol (TDG), a Schedule 2 compound[3] under

[1] Convention on the Prohibition of the Development, Production, Stockpiling and Use of Chemical Weapons and on Their Destruction. The treaty entered into force in 1997.

[2] The committee learned after report writing had been substantially completed that the operations start date for PCAPP had been reset to before the end of the calendar year 2015.

[3] Under the Chemical Weapons Convention, Annex on Chemicals, Schedule 1 chemicals are those that were developed, produced, and stockpiled or used as a chemical weapon, or are chemicals that would pose a high risk to the object and purpose of the Convention by virtue of their high potential for use as a chemical weapon. Schedule 2 chemicals pose a significant risk to the object and purpose of the Convention because they possess lethal or incapacitating toxicity as well as other properties that could enable them

TABLE 1-1 Inventory of Assembled Chemical Weapons at Pueblo Chemical Depot

Type of Munition	Size of Munition	Agent	Total No. of Munitions	Total Weight of Agent (lb)
M104 projectiles	155 mm	HD	33,062	386,820
M110 projectiles	155 mm	HD	266,492	3,117,960
M60 cartridges	105 mm	HD	383,418	1,138,760
M2 mortars	4.2 in.	HT	20,384	118,220
M2A1 mortars	4.2 in.	HD	76,722	460,340
Total			780,078	5,222,100

NOTE: HD, distilled mustard agent; HT, distilled mustard mixed with bis(2-chloroethylthioethyl) ether.

the CWC, the biodegradation process is also subject to CWC oversight. The TDG will be biodegraded within the ICBs.

The PCAPP process is described in Chapter 2 of this report. A subsequent report will address destruction of the stockpile at the BGAD facility, known as the Blue Grass Chemical Agent Destruction Pilot Plant (BGCAPP).

RELATED NRC 2013 REPORT

Treatment of the hydrolysate at PCAPP involves biotreatment followed by brine reduction and water recovery. The biotreatment system, including the BRS and a water recovery system (WRS), was extensively reviewed by an NRC committee in 2013 (NRC, 2013). Relevant findings and recommendations from that study are repeated in this report as necessary.

NEED FOR THE PRESENT NRC STUDY

Although biotreatment of toxic chemicals, brine reduction, and water recovery are established technologies, never before have these technologies been combined to treat mustard hydrolysate. And while extensive testing of these systems has been conducted by ACWA in a laboratory setting, the process has never been operated at full scale. The ACWA program managers and the PCAPP facility, including its contractor design, construction, and operations staffs, have every reason to believe that the facility will perform successfully. The NRC committee writing this report has reviewed extensive documentation on the PCAPP process and received two days of briefings from PCAPP staff and also from Colorado regulators, and it agrees that the PCAPP facility should be able to perform successfully. However, because it is a first-of-a-kind application of a combination of technologies to a unique matrix—mustard hydrolysate—there is still a possibility that the biotreatment, brine reduction, or water recovery processes may not perform satisfactorily.

to be used as a chemical weapon or as a precursor in one of the chemical reactions at the final stage of formation or in the production of a Schedule 1 chemical or another Schedule 2 chemical.

Finding 1-1. The ACWA program managers and the PCAPP facility, including its contractor design, construction, and operations staffs, have every reason to believe that the facility will perform successfully.

Finding 1-2. The NRC committee writing this report agrees that the PCAPP facility should be able to perform successfully. However, because the facility entails a first-of-a-kind application of a combination of technologies to a unique matrix—mustard hydrolysate—the committee also believes that there still exists a possibility that the biotreatment, brine reduction, or water recovery processes may not perform satisfactorily.

For example, the throughput of the ICBs may not match the throughput of the hydrolysate production. As another example, corrosion could lead to frequent repairs that shut down subsequent hydrolysate treatment. If these types of problems occur, then the hydrolysate may have to be collected and stored, and the actual munitions disassembly and mustard hydrolysate production may have to be interrupted until the downstream processes are able to catch up. Occurrences like these could also lead to other issues, such as storage of the hydrolysate for unanticipated long periods of time and idle periods for PCAPP workers.

Chemical weapons have been stored at the PCD for over 60 years, representing a steady-state or even increasing (due, for example, to the potential for the munitions to spring leaks) risk profile for the community; final destruction of these munitions at PCAPP will eliminate the primary risk posed by the munitions. Delays in the destruction process will halt this risk reduction and protract the risk that the community faces. Decisions that might delay PCAPP operations should consider potential impacts to the community, which in addition to protracting the risk associated with storing aging chemical weapons for longer periods of time, might include employee reassignment, furloughs, or even layoffs. Risk-based decision making should balance the new risks against the risk that has long been posed by the stockpile.

In the event that one or more of the hydrolysate treatment systems are shut down, even for a short time, destruction of the primary stockpile at PCD may need to be halted unless there is an alternative means for treating the hydrolysate. The committee firmly believes that destruction of the stockpile at PCD must continue, because it is destruction of the munitions and the agent that will eliminate the primary risk to the community. Hence, even though the PCAPP facility is expected to perform successfully, it is prudent, even necessary, to establish a backup plan—an alternative to the onsite treatment processes intended for the hydrolysate.

Finding 1-3. Destruction of the munitions and the agent will eliminate the primary risk to the local community. Hence, even though the PCAPP facility is expected to perform successfully, it will be prudent, even necessary, to establish a backup plan—an alternative to the onsite treatment processes intended for the hydrolysate.

One alternative for treatment of the hydrolysate that might be quickly implementable would be to ship the hydrolysate offsite to an existing, prequalified, and fully permitted treatment, storage, and disposal facility (TSDF). The offsite shipment would close the downstream portions of the plant that were constructed to treat the hydrolysate after its formation. Workers associated with these processes might need to be placed on furlough or laid off, which could cause them to seek alternative employment and PCAPP to lose trained staff. This is not the desire of ACWA program staff, the PCAPP contractors, or the local stakeholders.

To study this situation, the ACWA program requested the National Research Council (NRC) to form an ad hoc committee, the Committee on Review Criteria for Successful Treatment of Hydrolysate at the Pueblo and Blue Grass Chemical Agent Destruction Pilot Plants, to assess the process and the potential for offsite transport of the hydrolysate.

ORGANIZATION OF THIS REPORT

Chapter 2 provides information on the composition of the hydrolysate and describes the PCAPP processes for treating it in more detail. Chapter 3 discusses stakeholder concerns. Chapter 4 reviews regulatory considerations at the federal, state, and local levels and addresses requirements of the CWC. Chapter 5 discusses Department of Transportation regulations and identifies risks associated with the offsite shipment of hydrolysate. Chapter 6 establishes criteria for successfully treating the hydrolysate and identifies systemization data that should factor into the criteria and decision process for offsite transport and disposal of the hydrolysate. Chapter 7 discusses failure risks and contingency options as well as the downstream impacts of a decision to ship hydrolysate offsite.

There are also four appendixes. Appendix A contains the statement of task for the committee, Appendix B contains public interest and input documents received by the committee from the Colorado CAC, Appendix C provides biographical sketches of committee members, and Appendix D identifies the committee meetings and locations.

REFERENCE

NRC (National Research Council). 2013. *Review of Biotreatment, Water Recovery, and Brine Reduction Systems for the Pueblo Chemical Agent Destruction Pilot Plant.* Washington, D.C.: The National Academies Press.

2

Background

At the Pueblo Chemical Agent Destruction Pilot Plant (PCAPP), after the munitions have been robotically opened, the mustard agent will be destroyed by chemical hydrolysis with hot water (194°F), followed by addition of caustic to raise the pH to between 10 and 13. These processes yield a waste called "hydrolysate." The mustard agent's complete destruction is verified by sampling and analyzing the hydrolysate after neutralization. The hydrolysate is no longer chemical warfare material; the process also ensures that the product is not flammable. The hydrolysate will then be mineralized to water, carbon dioxide, and salts in the subsequent biotreatment step, using biodegradation and water recovery. This chapter provides an overview of the basic processes that will be used at PCAPP to destroy the munitions and mustard agent and provides information on the composition and concentrations of the hydrolysate. This is followed by descriptions of the downstream processing units, including the immobilized cell bioreactors (ICBs) and the brine reduction system (BRS).

BRIEF DESCRIPTION OF THE PCAPP PROCESS

Details of the process planned for destruction of the mustard agent at the Pueblo Chemical Depot can be found in prior National Research Council (NRC) reports (NRC, 2008, 2013). Only a brief overview is provided here, because the process design is not the focus of this report. A block diagram of the PCAPP process is presented in Figure 2-1. The approach to mustard agent destruction is divided into two broad phases. The first phase involves munitions processing and agent neutralization, which generates a secondary wastewater stream, called hydrolysate, that requires subsequent treatment. The steps in this first phase, which comprise the box labeled Munitions Processing and Agent Neutralization in Figure 2-1, include the following:

1. Remove munitions from the pallets.
2. Separate the propellant from the projectile.
3. Remove bursters at the linear projectile/mortar disassembly machine.
4. Send uncontaminated bursters and propellant offsite for destruction.
5. Wash out the mustard agent from the projectile bodies in the munition washout station (MWS) with a high-pressure water stream. The projectile bodies are then thermally decontaminated. The clean projectile bodies are shipped offsite for disposal.
6. Send mustard and washout water to the agent hydrolysis reactors, where the agent is hydrolyzed with hot water at 194°F. Caustic is added to raise the pH above 10 to prevent any reversible reactions back to mustard agent. The resulting product is called the agent hydrolysate.
7. Analyze the hydrolysate to verify that the mustard concentration is below 20 ppb and the distilled mustard (HT) concentration is below 200 ppb. The hydrolysate is then sent to one of three storage tanks. Each tank has a volume of 285,750 gal and provides up to 10 days of storage time, so the total storage period for hydrolysate is up to 30 days.
8. If the munition is leaking or contaminated with agent, the whole munition is destroyed by detonation using explosive destruction technology at an onsite facility separate from PCAPP.

The main chemical in the hydrolysate resulting from the mustard agent neutralization process is thiodiglycol (TDG). The liquid hydrolysate containing TDG will be treated in the second phase of the process by using immobilized cell bioreactors (ICBs) in the biotreatment area (BTA). The steps in this second phase include the following:

1. Transfer and treat the hydrolysate in the ICBs, where the main component, TDG, will be mineralized by the microorganisms under aerobic conditions. The hydrolysate is diluted and neutralized with H_2SO_4 in

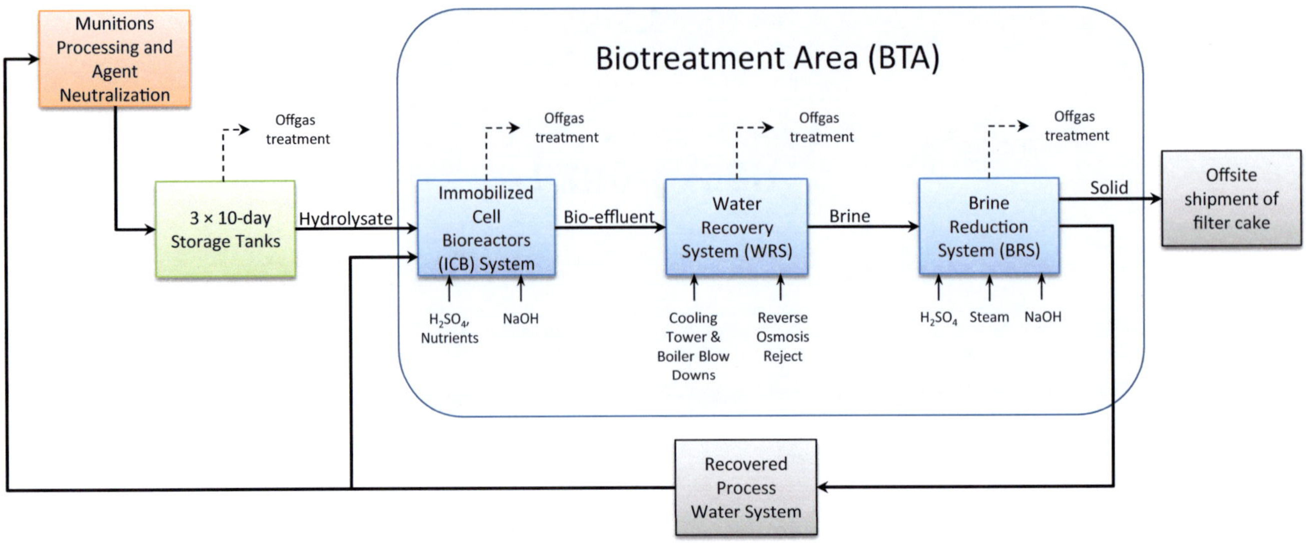

FIGURE 2-1 PCAPP biotreatment area process diagram. SOURCE: Adapted from Don Guzzetti, start-up field supervisor, PCAPP, and Dave DeLesdernier, support, Battelle, "Biotreatment Area Risk Reduction & Mitigation," presentation to the committee on July 29, 2014.

the ICB's feed tanks prior to entering the ICBs. This step is represented by the box labeled Immobilized Cell Bioreactors System in Figure 2-1. A removal efficiency of at least 95 percent of the influent TDG is the treatment goal for the ICBs. There is flexibility in this value, because failure to attain this TDG removal efficiency does not automatically trigger offsite disposal, as described more fully in Chapter 7 (section Impacts If ICB Systems Underperform or Do Not Perform).

2. The liquid effluent from the ICBs goes to the water recovery system (WRS), which is a storage and mixing tank, and is then processed by the BRS. This step is represented by the box labeled Water Recovery System in Figure 2-1.
3. The BRS consists of an evaporator to yield a brine and biomass concentrate, which is sent to a crystallizer for further solids concentration. The water product from both the evaporator and crystallizer flows through a column of granular activated carbon for removal of trace organic constituents. The principal purpose of the BRS is to recover solids and generate a water stream that is of sufficient quality to become recycled process water. This step is represented by the box labeled Brine Reduction System in Figure 2-1.
4. The salts and biomass solids recovered from the crystallizer are dewatered in a filter press and sent offsite for disposal.
5. The recovered water is returned to a process water storage tank.
6. The offgas from several of the treatment units can contain traces of toxic volatile substances and/or odorous compounds, as shown in Figure 2-1. For example, the volatile organic compounds (e.g., vinyl chloride) in the feed to the ICBs will be stripped from the bioreactor solution and appear in the offgas from the ICBs. These offgases are collected and passed through granular activated carbon adsorbers prior to release to the atmosphere.

PRODUCTION AND CHARACTERIZATION OF HYDROLYSATE

The hydrolysate will be generated in batches using two reactors, each having a working capacity of 3,600 gal. The maximum production rate of the liquid hydrolysate is expected to be 10,200 lb/hr.[1] The peak liquid flow rate of hydrolysate to the three 10-day storage tanks is expected to be 19 gal/min, or 27,360 gal/day.

Through the Assembled Chemical Weapons Alternatives (ACWA) program, biodegradation testing of hydrolysate was conducted by Guelta and Fazekas-Carey in 2003. The test hydrolysate for the 2003 studies was generated from drained liquid agent and the solid heel material in 4.2-in. mortar rounds containing distilled mustard (HD) stored at the Pueblo Chemical Depot. The components and concentrations detected in the hydrolysate for these 2003 studies are given in Table 2-1. The Resource Conservaton and Recovery Act (RCRA) permit is another source for the composition of the PCAPP agent hydrolysate (third column in Table 2-1). The characterization information provided in Table 2-1 is considered representative of what will be encountered when the PCAPP facility begins processing munitions. The principal

[1] Bill Steedman, senior process engineer, PCAPP, "PCAPP Agent Treatment Process," presentation to the committee on July 29, 2014.

TABLE 2-1 HD Hydrolysate Characterization from 2003 Biotreatment Testing and RCRA RD&D Permit Waste Analysis Plan (milligrams per liter unless otherwise noted)

Component	Concentration from Guelta and Fazekas-Carey (2003)	Anticipated Agent Hydrolysate Composition
Thiodiglycol (TDG)	17,537	52,250
Dithiane	2,093	1,370
Thioxane	47.9	N.R.
Chemical oxygen demand (COD)	43,100	N.R.
Total organic carbon (TOC)	8,120	N.R.
Percent TOC as TDG (%)	84.9	N.R.
COD:TOC ratio	5.31	N.R.
Sulfate	84	N.R.
Sulfur	6,010	N.R.
Total dissolved solids (TDS)	28,000	N.R.
Total suspended solids (TSS)	1,000	8,676
pH	13	10 to 12
Specific gravity	1.03 g/mL	1.066 g/mL
Aluminum	1.99	N.R.
Ammonia	N.R.	2.99
Arsenic	0.579	2.02
Barium	0.033	2.47
Cadmium	3.2	0.0
Calcium	10.9	N.R.
Chloride	10,800	N.R.
Chromium(VI)	N.R.	0.90
Chromium (total)	N.R.	1.64
Copper	0.281	6.52
Fluoride	N.R.	0.54
Iron	520	2,160
Lead	3.69	1.43
Magnesium	5.74	N.R.
Mercury	0.013	0.154
Molybdenum	0.065	N.R.
Nickel	0.330	1.754
Phosphorus	0.456	N.R.
Potassium	15.2	N.R.
Silver	5.73	N.R.
Sodium	10,630	N.R.
Zinc	3.59	3.81
1,2,3-Trimethylbenzene	N.R.	0.0
1,2,4-Trimethylbenzene	N.R.	0.0
1,2-Dichloroethane	N.R.	181
2-Butanone	N.R.	2.56
2-Hexanone	N.R.	6.20
4-Methyl-2-pentanone	N.R.	7.30
Acetone	N.R.	6.29
Acetonitrile	N.R.	0.0
Benzene	N.R.	0.319
Chloroform	N.R.	0.319
Diethyl ether	N.R.	0.288
Hexane	N.R.	0.0
Methylene chloride	N.R.	1.31
m- and *p*-Xylenes	N.R.	0.07
Naphthalene	N.R.	0.038
Tetrachloroethene	N.R.	8.67
Toluene	N.R.	0.066
Trichloroethene	N.R.	1.52
Vinyl chloride	N.R.	11.9
2-Methylphenol	N.R.	0.0
2-Nitrophenol	N.R.	0.0
3- and 4-Methylphenol	N.R.	0.0
1,4-Oxathiane	N.R.	534

NOTE: N.R., Not reported.

SOURCE: PCAPP (2006), Table C-9. Concentrations in Table C-9 are based on the treatment of distilled mustard agent at the Aberdeen Chemical Agent Disposal Facility (ABCDF).

components of hydrolysate are TDG, other organic compounds (mainly dithiane and thioxane), suspended solids, and dissolved inorganic compounds (predominantly sodium chloride and iron salts). Many trace organic and inorganic compounds are also detected.

Another indicator for the character of the agent hydrolysate is obtained from the reported design parameters for the ICBs (Table 2-2). The plan calls for diluting the agent hydrolysate after it leaves one of the 10-day storage tanks, with the recycled process water in the ratio of one part hydrolysate to seven parts process water before it enters the feed tank to the ICBs. The dilution is required to maintain the influent TDG concentration at a level (7,000 mg/L) that results in nontoxic levels for the bacteria within the ICBs. The toxic threshold for TDG to the biomass is estimated to be 2,000 mg/L. This means that TDG concentrations must be maintained below 2,000 mg/L in the ICB units. The treatment goal (RCRA RD&D permit performance) is to remove at least 95 percent of the TDG in the ICBs,[2] so that steady-state TDG biodegradation should keep the reactor contents below the toxic threshold concentration. Furthermore, the hydraulics within an ICB chamber approaches complete mix from the coarse bubble aeration system. This provides for significant dilution of the influent feed TDG concentration. It is expected that a feed TDG concentration of 7,000 mg/L or less can be supplied to the ICBs during steady-state operation to prevent toxicity to the bacteria from the TDG.

TABLE 2-2 Key Design Operating and Feed Characteristics for the Immobilized Cell Bioreactor Units

Design Characteristic	Hydrolysate (total)	ICB Influent (per unit)[a]
Flow (gal/day)	16,766	8,383
Hydraulic retention time (days)		4.98[b]
Volume (gal)		41,783
Concentration of TDG (mg/L)	56,000	7,000
Concentration of TSS (mg/L)	8,000	1,000
Concentration of COD (mg/L)	120,000	15,000
Concentration of TOC (mg/L)	26,400	3,300
Concentration of iron (mg/L)	2,160	270
Concentration of NaCl (mg/L)	57,600	7,200
pH	10-13	7-8[c]

[a] The ICB influent is diluted with process water (1:7), resulting in a total flow of 134,128 gal/day across 16 ICB units.

[b] This was calculated from the flow and volume shown.

[c] This will be maintained via acid production within the unit and caustic or acid addition as needed.

SOURCE: Paul Usinowicz, PCAPP technical advisor, and Yakup Nurdogan, PCAPP lead industrial wastewater treatment engineer, "Biotreatment Area (BTA) Cradle to Grave," presentation to the committee on July 30, 2013, and NRC (2013).

DESCRIPTION OF THE IMMOBILIZED CELL BIOREACTORS

The TDG and other compounds in the agent hydrolysate will be aerobically biodegraded in an ICB, where the biomass is attached to a carrier medium. The bioreactor system consists of 16 ICB units configured in four parallel modules with four units per module. Each ICB unit has a volume of about 42,000 gal and is divided into three chambers in series with volumes of 21,000 gal, 10,500 gal, and 10,500 gal, respectively. The hydrolysate feed to each ICB unit requires dilution (one part hydrolysate to seven parts process water) to achieve nontoxic levels of the TDG in the bioreactor. The expected hydraulic loading rates to each ICB unit are anticipated to vary between 4,800 and 9,700 gal/day (BPT, 2006), yielding an average hydraulic retention time of between 8.6 and 4.3 days, respectively. The ICBs are expected to remove at least 95 percent of the influent TDG.

Additional details of the physical layout and planned operation of the ICBs are provided in a recent NRC report (2013). The committee that prepared the 2013 report reviewed the bench and pilot-scale studies that evaluated biodegradation of TDG and the planned design and operating procedures for the ICBs. Several issues pertinent to the operation of the ICBs were identified that merit close attention, and the major ones are summarized in Table 2-3.

[2] PCAPP, RCRA RD&D Stage III, Class 3, Permit Modification Request, Revision 0, November 2006, Appendix D, Waste Analysis Plan, C-2a(3) (d), p. C-11.

DESCRIPTION OF THE BRINE REDUCTION SYSTEM

The BRS is designed to recover water from a number of sources within the PCAPP system, provide a recycle water stream for the process water needs of the facility, and allow the PCAPP system to achieve zero liquid discharge operation. The salts and biomass recovered from the crystallizer will be dewatered in a filter press and sent offsite for disposal as a solids residue. The feed for the BRS is sourced from the WRS, where a number of streams, including the effluent from the biotreatment system, are collected for feed to the BRS. For the purposes of this discussion the WRS, which is an aerated tank system, will be considered as part of the BRS. This section provides a simple description of the BRS. A more detailed description is available in previous reports (NRC, 2013; PCAPP, 2014).

The major equipment in the BRS consists of WRS tanks, an evaporator system, a crystallizer system, and a granular activated carbon system. The WRS tanks provide an equalized feed to the evaporator system, where the water stream is pretreated to remove carbon dioxide through acidification and aeration before concentrating the brine stream through water evaporation and distillation to recover water for recycle. The recovered water stream is processed through

TABLE 2-3 Issues Identified in a Review of the Immobilized Cell Bioreactors and Biotreatment of the Hydrolysate

Issue	Explanation
Toxicity	At elevated concentrations (>2,000 mg/L), TDG was observed to become inhibitory in a previous biodegradation test. Once the ICBs are in a steady state, the 1:8 dilution of the agent hydrolysate and treatment goal of at least 95 percent destruction of TDG should maintain TDG concentrations well below toxic concentrations in the ICBs. Toxic inhibition may be a concern during start-up and off-normal operation, especially if there is elevated loading of the TDG or periods with poor biodegradation of the TDG.
Inhibition by heavy metals	The hydrolysate contains some heavy metals that can exert antimicrobial properties, such as Ag and Cu. The levels of metals expected is not likely to be a major concern during steady-state biotreatment, but toxicity from heavy metals should be explored if performance of the ICBs is below expectations.
Inhibition by sulfide	The organic sulfur in TDG is converted to sulfate during the aerobic bio-oxidation of TDG. If the conditions in the bioreactor and biomass are not fully aerobic, anaerobic microniches may occur within the immobilized biomass and facilitate the reduction of sulfate to sulfide. Sulfide can be toxic to bacteria at concentrations >100 mg/L. Sulfide toxicity can be minimized with sufficient aeration and development of fully aerobic biofilms.
pH	The TDG biodegradation tests suggest an optimum pH range of 7 to 8 for the ICBs. The hydrolysate with high pH is first neutralized with H_2SO_4 in the feed tank prior to entering the ICBs. TDG bio-oxidation within the ICBs generates sulfuric acid. Therefore, pH control is essential, and it may be challenging to maintain the proper pH.
Temperature	The past biodegradation testing with TDG observed stable performance in the temperature range of 46°F to 95ºF. Steam injection can be used to heat the ICBs, but there are no provisions for cooling the ICBs. It is not known if summertime heat and solar load can cause the temperatures in the ICBs to exceed the optimum range.
Solids buildup	The feed hydrolysate to the ICBs is likely to contain significant TSS. The feed also contains iron, and phosphate is added as a nutrient. These inorganics are likely to precipitate and produce inorganic solids. Biomass, a third important solid in the ICBs, may build up excessively and alter the hydraulic characteristics. Periodic sloughing of the biomass could contribute to intermittent releases of large quantities of biomass solids into the effluent from the ICBs, resulting in a higher solids loading to the BRS.
Oxygen demand and flux issues	Air will be supplied to the ICBs via coarse diffusers to provide oxygen for the biodegradation of TDG. The air supply is evenly distributed throughout the three chambers, with all of the TDG loading entering in the first chamber. Consequently, the demand for oxygen in the first chamber may exceed supply, resulting in lack of oxygen within the biomass/foam medium.
Start-up issues	Given the factors identified above with the unique and complex hydrolysate feed, the start-up of the ICBs is likely to be a critical period, especially considering the potential for toxic inhibition and solids buildup and challenges with the air supply and maintenance of the optimum pH.
Characterization and monitoring	Little is known about the composition and variability of the ICB effluent, especially in terms of residual nonbiodegradable compounds and soluble microbial products. Similarly, little is known about the residual volatile organic compounds in the offgas. During initial operation, the composition and characteristics of the ICB effluent and offgas should be monitored to anticipate potential long-term concerns for downstream processing. For example, the presence of uncharacterized compounds in the hydrolysate feed to the ICBs may build up in the recycled process water.

SOURCE: NRC (2013).

TABLE 2-4 Issues Identified in a Review of the Brine Reduction System

Issue	Explanation
Carbon adsorbers meeting permit requirements with acceptable pressure drop	There is uncertainty about the composition of the water leaving the carbon adsorbers and whether there will be plugging that leads to higher-than-anticipated pressure drop across the bed. It may be necessary to replace the carbon beds and backwash them more frequently than currently anticipated.
Quality of distillate from the crystallizer	There is uncertainty about the concentrations of the organic compounds in the distillate from the crystallizer. If they are too high, it may be necessary to improve de-entrainment in the crystallizer.
Too much water in filter cake	The presence of organic matter in the filter cake may make it very difficult to reach the desired solids content. Methods for additional dewatering or alternate disposal options would then be needed.
No prior full-scale testing data available	The amounts and identities of all of the components in the BRS feed stream are not known. PCAPP needs to monitor the BRS feed and effluent streams during hydrolysate treatment and be prepared for unanticipated components or concentrations.
Acid hydrolysis of biomass	If biomass components hydrolyze during acid treatment, there could be operational problems downstream. Installing a clarifier between the ICBs and the BRS is a potential solution.

SOURCE: NRC (2013).

carbon beds and then forwarded to the process water system. The concentrated brine is further processed in the crystallizer system. As with the evaporator system, the distillate water is fed to carbon adsorption and then to the process water system. The concentrated brine slurry is sent to a belt filter system. The filtrate from the belt filter press is recycled to the front of the crystallizer. The filter cake from the filter system is packaged and sent offsite for disposal.

A number of risk reduction and minimization studies are planned prior to the start-up of munitions processing at PCAPP, to be conducted in tandem with systemization. The most significant of these tests is the operation of the units with feed from the TDG surrogate testing of the ICBs. This test will provide a feed that mimics the expected operation as closely as possible in that it will have a brine feed with a salt and biosolid content that will be similar to what is expected during operation. In that respect it is foreseen that it will be possible to get an early indication of the water quality, fouling characteristics, and other processing capabilities of the operating units.

Just as Table 2-3 identifies issues that may be of concern with the ICBs, Table 2-4 identifies such issues for the BRS. These items are taken from the 2013 NRC report.

REFERENCES

BPT (Bechtel Pueblo Team). 2006. *System Design Description (SDD) for Biotreatment System No. B09.* Pueblo, Colo.: Bechtel.

Guelta, M.A., and L. Fazekas-Carey. 2003. *Biodegradation of Hydrolyzed Mustard from an Assembled Chemical Weapons Assessment (ACWA) Projectile Washout Study.* ECBC-TR-291. Aberdeen Proving Ground, Md.: Edgewood Chemical Biological Center.

NRC (National Research Council). 2008. *Review of Secondary Waste Disposal Planning for the Blue Grass and Pueblo Chemical Agent Destruction Pilot Plants.* Washington, D.C.: The National Academies Press.

NRC. 2013. *Review of Biotreatment, Water Recovery, and Brine Reduction Systems for the Pueblo Chemical Agent Destruction Pilot Plant.* Washington, D.C.: The National Academies Press.

PCAPP (Pueblo Chemical Agent Destruction Pilot Plant). 2006. PCAPP RCRA RD&D Stage III, Class 3, Permit Modification Request, Revision 0. Attachment D, Waste Analysis Plan. Pueblo, Colo.: Pueblo Chemical Agent Destruction Pilot Plant.

PCAPP. 2008. RCRA Permit Attachment D Waste Analysis Plan, Modification #6, Pueblo Chemical Agent Destruction Pilot Plant.

PCAPP. 2014. *White Paper: Bio-Treatment Area Risk Reduction and Mitigation.* 24852-30H-BTA-V0001. Pueblo, Colo.: Pueblo Chemical Agent Destruction Pilot Plant.

3

Stakeholder Interests and Issues

BACKGROUND

A variety of stakeholders, including the Assembled Chemical Weapons Alternatives (ACWA) program, the Pueblo Chemical Agent Destruction Pilot Plant (PCAPP), regulators, members of the Colorado Citizens' Advisory Commission (CAC), and other local entities, are interested in and affected by activities at the PCAPP facility. The discussion in this chapter, however, focuses on the interests and input of local members of the general public, relying primarily on the CAC as the institutional representative of the local and state populations.

The public involvement process at PCAPP is well established and uses a variety of approaches to keep the nearby communities informed and involved in project activities and decisions. As described in *Review of Secondary Waste Disposal Planning for the Blue Grass and Pueblo Chemical Agent Destruction Pilot Plants* (NRC, 2008), the involvement process includes the following:

- An Outreach Office that maintains a web site and a mailing list, now estimated at 1800 recipients, that distributes fact sheets, newsletters and project updates; runs a speakers' bureau that makes presentations to businesses, local officials, schools and civic organizations and minority communities; and supports the local reuse authority which is planning for reuse of the property after completion of demilitarization.
- Frequent interaction with elected officials, and staff from the regulatory and emergency management agencies.
- A Citizens' Advisory Commission (CAC), formed as a consultative body to receive citizen and state concerns. It is comprised of nine members appointed by the state governor,[1] and is administered by the Colorado Department of Public Health and Environment. The CAC, which meets 10 times a year, is the focal point for public discussion of PCAPP issues.

In addition, since 2010, the Program Executive Office (PEO) for ACWA has periodically posted on Twitter and Facebook, providing information about operations and programs to the broader public (it calls itself the "official source of PEO ACWA's chemical weapons destruction news").

METHODOLOGY

The committee adopted a three-pronged approach in seeking input from local stakeholders into the criteria and decision points to be considered in completing the current task. This approach was discussed and suggestions for enhancing public input were solicited in public presentations, shown in Appendix B, Exhibit 1, both of which took place in Pueblo in July 2014. Input from local stakeholders was sought through

- Open discussion at the July NRC and CAC meetings.
- Establishment of a dedicated e-mail address to which comments could be submitted: Comments_for_NRC_Hydrolysate_Committee@nas.edu. This address was established in mid-July, prior to meeting with the CAC. It was printed on business cards that were distributed at the CAC meeting, made available at the Outreach Office, and publicized in an article in the local newspaper, *The Pueblo Chieftain* (see Appendix B, Exhibit 2).
- A webinar combining written and spoken input to clarify and further discuss outstanding issues, if needed.

The CAC had been briefed by the Army and given an opportunity to comment on the statement of task prior to its issuance. The Hydrolysate Committee held its first meeting in Pueblo in July 2014, coincident with that month's meeting of the CAC to facilitate local attendance, and it invited CAC members to join it for the presentations and open discussion. Two members of the CAC, including the chair, attended the 2-day open meetings, and NRC committee members

[1] 50 USC 1521 (m).

attended the CAC meeting to introduce themselves, provide an overview of the study and respond to questions, and further emphasize the importance of community input.

Interests and Input of the CAC

There is common agreement among all stakeholders on the goals of avoiding unnecessary delay in destruction of the stockpile, reducing risk and assuring the safety of workers and the community, and using resources and funding efficiently. However, the priority placed on these objectives differs across stakeholders, in particular between the CAC and the other stakeholders.

For example, the CAC, which is strongly committed to onsite biotreatment, has consistently opposed offsite shipment of hydrolysate. Although the views of the broader public are less well researched and known, the CAC's opposition to offsite shipment and mistrust of the Army has a long and well-documented history that was further increased by its experience of the way shipments at Newport were conducted (see especially Noblis, 2008; NRC, 2008). As the discussions at the July Pueblo meeting clearly showed, this opposition continues and is permeated by a skepticism that the current NRC study represents, in the words of one CAC member in informal conversation, "yet another" attempt by "the folks in the Pentagon" to negate the Army's commitment to the neutralization/biotreatment package that was selected, with strong involvement and support, by the Pueblo and Blue Grass communities and embodied in the 2003 ACWA legislation. The CAC chair also noted an apparent disconnect between ACWA's public declarations of the safety of biotreatment and the current effort to consider the technology's possible failure.

A comparison of the CAC statement presented by the CAC chair at the NRC Hydrolysate Committee meeting (see Appendix B, Exhibit 3) and the statement provided to the NRC in 2008 that was developed jointly by the Pueblo and Blue Grass CACs shows fundamentally similar concerns about offsite shipments. In addition to the loss of trust if ACWA were to rescind its agreement to treat hydrolysate onsite, specific concerns itemized in the 2014 statement that find a parallel in the 2008 statement include these:

- Violation of environmental justice principles, specifically, that each community should take care of its own waste and not dump it on someone else.
- Political opposition and potential litigation from host communities and communities along transportation routes.
- Transportation risks, including the cost of spills and emergency response assistance, to communities along the routes; the efficiency and thoroughness of the coordination of notification, monitoring, and management of shipments.
- Identification of transportation routes and the need for additional National Environmental Policy Act of 1969 (NEPA) review and regulatory changes that would affect cost and schedule, including compliance with the Chemical Weapons Convention.
- The compliance record of selected treatment, storage, and disposal facilities (TSDF) (relating to both the Occupational Health and Safety Administration and environmental justice compliance) and assurance of timely and fiscally responsible performance.
- Potential loss of jobs and economic opportunities for the Pueblo community.
- Loss of water sourced from the Pueblo facility if hydrolysate is shipped rather than treated onsite (this concern had not previously been listed).

In addition, the CAC chair strongly recommended that stakeholders be included in establishing the criteria and decision points used to determine that offsite shipment of hydrolysate is the only viable option, as well as in the decision-making process should such shipment need to be addressed. She reiterated that shipment should not be considered as the first and only remedy but as the remedy of last resort. She also recommended consideration of cost and safety and the need for continuous communication with the community and the Colorado Department of Public Health and Environment (CDPHE) to avoid misunderstanding and to facilitate the permitting changes that would be needed if a decision is made to ship offsite.

Subsequent input provided at the CAC meeting highlighted members' continuing commitment to deployment of the selected technology of biotreatment. For example, at the CAC meeting, the CAC chair expressed the hope that the NRC study would be placed on a shelf and not actually be used. She emphasized that from the CAC's perspective, if problems arise, hydrolysate shipment is "not the first choice" but "the very last, the ultimate last, the nothing-else-can-be-done last" choice. She also expressed the desire that the NRC Hydrolysate Committee's findings be clearly conveyed in a publicly accessible forum upon completion of the report.

Similarly, the follow-up CAC written statement of concerns (see Appendix B, Exhibit 4) reemphasized members' commitment to biotreatment and their confidence in PCAPP personnel to successfully mitigate and resolve potential limitations that could arise. In its statement, the CAC expressed the opinion that total failure of the biomass, including failure of the biomass seed to acclimate to the hydrolysate feed, or failure to reduce the thiodiglycol to required levels would represent a failure of the system. The CAC listed a series of possible problems and reiterated its confidence in ACWA and PCAPP personnel to address them. The CAC also noted that it was not just "a simple matter of shutting down the biotreatment area (BTA) should one or all parts of the BTA fail," but that offsite shipment would incur additional costs as well as loss of schedule. The CAC cited costs and schedule issues, which included changes to NEPA and permitting; construction costs of building the loading dock, piping, and road access; increased water costs arising from loss of recycled water; coordination among federal, state, and local agencies; and negotiations with the Organisation for the Prohibition of Chemical Weapons (United Nations) (OPCW). In addition, the CAC recommended an extensive educational campaign to address the skepticism that would arise among members of the local community, who had long accepted the "total package" of neutralization followed by biotreatment.

Interests and Input of Other Local Stakeholders

Only one comment was received by the NRC committee at the dedicated e-mail address that provided a perspective different from the CAC's. The comment, which was submitted by Carl Ballinger, Chemical Stockpile Emergency Preparedness Program (CSEPP), Federal Emergency Management Agency (FEMA) Coordinator for Pueblo County, offered strong support for deploying offsite hydrolysate shipments should they be needed:

> I have been involved in chemical weapons storage and the Chemical Stockpile Emergency Preparedness Program for over 24 years; first as a Chemical Corps soldier assigned to the then Pueblo Depot Activity, and now as the CSEPP Coordinator for Pueblo County. I think it is important to remember what the primary mission is: safely destroying the stockpile. Anything that hinders or slows down that process needs to be addressed quickly and effectively, regardless what the cause. If the delay is caused by the inability of the biotreatment to function effectively then we need to move quickly to consider other options, to include offsite shipment.

Other than the above comment, the NRC received no comments from local stakeholders through the dedicated e-mail address. Discussion during a conference call with staff from the PCAPP public affairs office, the CDPHE, and the CAC indicated that chemical materiel activities at PCAPP were not a significant concern.[2] The public affairs officer noted that interest levels varied over time, according to activity levels at the facility. Currently, the program is said to be in a lull, a conclusion supported by data collected by CSEPP for a 2013 Web-based survey. The survey showed that the percentage of residents aware of chemical agents at the Pueblo Chemical Depot decreased by 7 percent between January 2004 and January 2013, from 93 percent to 86 percent (FEMA, July 2013). However, those data also demonstrate that local residents have a relatively high level of awareness (86 percent for local residents and 84 percent for local business representatives), and PCAPP public affairs staff expect to see an increase in interest and queries as PCAPP operations begin. Although attendance by local stakeholders at CAC meetings typically is sparse, participants on the conference call agreed that the general public nevertheless relies on the CAC to act as a watchdog for the community's interests. Moreover, they believed a decision by ACWA to initiate offsite shipment of hydrolysate and rescind its commitment to the accepted neutralization/biotreatment package would immediately spark interest and controversy among local, and even national, public stakeholders. And, as clearly demonstrated by the history of the stockpile program, such controversy could lead to conflict, litigation, and unacceptable delay in the ACWA program. Conversely, cooperation among stakeholders could facilitate achievement of program goals.

[2] September 5, 2014, conference call with Judith Bradbury and Hank Jenkins-Smith, members, Hydrolysate Committee; Irene Kornelly, chair, Colorado CAC; Jeannine Natterman, CDPHE; John Norton, member, Colorado CAC; Sandy Romero, PCAPP communications manager; Nancy Schulte, study director, Hydrolysate Committee; and Thomas Schultz, public affairs specialist, PCAPP.

SUMMARY

As demonstrated in the following technical chapters, there is strong confidence in the success of the chosen technologies at PCAPP. However, as with any new application, unexpected issues might arise and appropriate mitigation measures might be required, often at short notice. In view of the established role of the CAC in the community, in combination with members' strong commitment to biotreatment, lingering concerns about the equivalent strength of ACWA's commitment, and skepticism about any suggestion to institute offsite shipment, many opportunities exist for misinterpretation of motives and technical decisions that are made as preparatory work continues. It will therefore be critical for ACWA to initiate discussions with the CAC well before operations begin and to put in place an institutional mechanism that can focus on bridging any disconnect between technical priorities and the community's social priorities. Such a mechanism would be supplementary to, yet part of, the CAC and would build on the sound working relationships that PCAPP and the CAC have been developing over many years. It would aim to ensure regular, open communication throughout operations, enhancing the probability that program goals can be achieved despite differing priorities.

FINDINGS AND RECOMMENDATIONS

Finding 3-1. All stakeholders (ACWA, PCAPP, state regulators, CAC, and other local stakeholders) agree on the need for (1) avoiding unnecessary delay in the efficacious destruction of the stockpile; (2) reducing risk and ensuring the safety of workers and the community; and (3) efficient use of resources and funding. However, the priority placed on these objectives differs across stakeholders.

Finding 3-2. Given the uncertainty about technical issues that could arise during operations, it is critical to avoid surprises presented by decisions that may be needed, especially those relating to offsite shipment of hydrolysate.

Finding 3-3. There is a history of distrust and a continuing lack of confidence on the part of some local stakeholders that offsite shipment will occur only as a last resort.

Finding 3-4. Disagreement about the need for offsite shipment could lead to conflict, litigation, and unacceptable delay in the ACWA program. Conversely, cooperation among stakeholders could facilitate the achievement of program goals.

Finding 3-5. Because of the potential for technical uncertainties in a context of different priorities and a history of distrust, it is critical that all stakeholders receive a credible commitment from ACWA that they, the stakeholders, will be given a clearly defined and meaningful role in any deliberations that could lead to a decision to ship offsite.

Finding 3-6. PCAPP staff have developed a sound working relationship with the CAC and local stakeholders that serves as a foundation for establishing a credible consultation process.

Recommendation 3-1. In consultation with the CAC, ACWA should institutionalize an explicit consultation process that focuses on the potential for offsite shipment. This process should be established immediately and give stakeholders a clearly defined and meaningful role. The consultation process should (1) be supplementary to the more general role of the CAC; (2) provide to the CAC regular updates on the status of operations as they bear on the possible need for offsite shipments; and (3) be explicitly designed to ensure there are no surprises on the part of stakeholders if they are called on to consider offsite shipments.

Finding 3-7. In order to avoid delay and the imposition of undue risk on workers and the local community, advance preparations for offsite shipment may need to be initiated at least several months before a decision is made to actually initiate offsite shipments. Advance preparations will need to be conducted as part of the stakeholder consultation process so that stakeholders are informed as these preparations are in progress.

Finding 3-8. Absent stakeholder consultation and understanding, these preparations could be subject to misinterpretation.

Recommendation 3-2. Once a process is in place for stakeholder consultation in the determination of the need for offsite shipments, advance planning on associated regulatory and plant issues for such shipment should be expedited.

REFERENCES

FEMA (Federal Emergency Management Agency). 2013. Comment sent to committee by Carl Ballinger via e-mail. July.

Noblis. 2008. *Offsite Disposal of ACWA Hydrolysates*. NTR 2008-61129. Falls Church, Va.

NRC (National Research Council). 2008. *Review of Secondary Waste Disposal Planning for the Blue Grass and Pueblo Chemical Agent Destruction Pilot Plants.* Washington, D.C.: The National Academies Press.

4

Regulatory Requirements for Offsite Hydrolysate Shipment and Treatment

INTRODUCTION

The primary mission of the Pueblo Chemical Agent Destruction Pilot Plant (PCAPP) is to safely destroy the Pueblo Chemical Depot chemical stockpile. Should it become necessary for PCAPP to send hydrolysate offsite in order to continue destruction of the chemical stockpile, all regulatory requirements must be identified and prebriefed with the Colorado Department of Public Health and Environment (CDPHE) and any other relevant regulatory bodies and must be coordinated with the Citizens' Advisory Commission (CAC), Assembled Chemical Weapons Alternatives (ACWA), and other stakeholders so that the destruction mission is not unduly delayed.

RCRA PERMITTING

Regulatory Background

The Resource Conservation and Recovery Act (RCRA) sets up a program for cradle-to-grave hazardous waste management.[1] Treatment, storage, and disposal facilities (TSDFs) are required to obtain permits that establish specific operating conditions as well as the requirements for facility closure.

While RCRA is a federal program administered by the U.S. Environmental Protection Agency (EPA), states may seek authorization from EPA to operate their own programs, and most states have this authorization. The Colorado Department of Public Health and Environment (CDPHE) is authorized to administer the RCRA program within the state of Colorado. CDPHE's RCRA program is defined in the Code of Colorado Regulations (CCR) at 6 CCR 1007-3.

[1] 42 U.S. Code §6901 et seq.; Code of Federal Regulations (CFR) at 40 CFR §§260 to 272.

Hazardous Waste Listings and Characteristics

Having received authorization from EPA to administer its own RCRA program, a state can make its regulatory schemes more stringent or broader in scope than the federal program. Colorado, like many of the states with past or present chemical demilitarization facilities, has chosen to make its program broader in scope than the federal RCRA program by listing specific chemical agent wastes as hazardous. Colorado-specific hazardous wastes listings include chemical munitions containing the following agents: Levinstein mustard agent (H), distilled mustard agent (HD), mustard gas, distilled mustard mixed with bis(2-chloroethylthioethyl) ether (HT), and sarin (GB) (waste codes P909, P910, and P911) and waste chemical weapons (waste codes K901, K902, and K903).[2]

In 2012, Colorado added hydrolysate from the treatment of 105- and 155-mm howitzer munitions and 4.2-in. mortar munitions as a waste chemical weapons listed waste and assigned to it waste code K903. The Colorado regulations define K903 hydrolysate as waste generated from the chemical neutralization of mustard agent by the addition of water and subsequent manipulation to a sustained and stable $pH > 10$ to ensure destruction of sulfonium ions and thiodiglycol (TDG)-mustard aggregates. K903 is not an acute listed waste but only a toxic listed waste exhibiting the RCRA toxicity characteristic. The other waste chemical weapons are listed as acute hazard waste, not only for exhibiting the toxicity characteristic but also for their reactivity and corrosivity.[3]

The RCRA program also establishes RCRA characteristics, which include ignitability, corrosivity, reactivity, and toxicity. The characteristics entail specific tests or definitions

[2] See 6 CCR 1007-3, Part 261.

[3] Solid and Hazardous Waste Commission/Hazardous Materials and Waste Management Division, 6 CCR 1007-3, Hazardous Waste, Addition of K903 (Hydrolysate) Listing, adopted by the commission on November 20, 2012.

that, if met, would identify the waste as a "characteristic hazardous waste." Hence, to be hazardous, wastes can either be a listed hazardous waste or a characteristic hazardous waste, or both (i.e., a listed waste that also meets one or more characteristics).

The means by which a waste is defined as hazardous is important because if a waste is a listed waste, any residue resulting from treatment of the waste is also defined as that listed waste (the derived-from rule), even if it no longer contains those attributes that made it a listed hazardous waste in the first place. For PCAPP, this means that the hydrolysate and all downstream secondary wastes retain the K903 listing designation, including the filter cake resulting from the biotreatment process. RCRA provides a delisting process that is available for waste that the waste-generating entity believes no longer meets the listing description. Delisting, however, can be a long and arduous process.

In contrast, if a waste is hazardous only because it demonstrates one or more of the RCRA characteristics, once that waste is treated or otherwise managed so that it no longer demonstrates the characteristic, it becomes a nonhazardous waste, subject to the state program for solid (nonhazardous) waste management.

RCRA Land Disposal Restrictions

The RCRA Land Disposal Restrictions (LDRs) establish treatment standards that hazardous wastes must meet before they can be land-disposed. The LDRs apply to both listed and characteristic wastes and entail specified treatment technologies or constituent concentrations that must be met. The federal program does not identify chemical weapon or agent wastes as listed hazardous wastes, so the federal LDRs do not apply to PCAPP hydrolysate (Colorado K903 waste) unless it exhibits one or more of the RCRA characteristics. However, Colorado recently established state-specific LDR treatment standards for K901, K902, and K903.[4] These LDR treatment standards would only apply to treatment of hydrolysate at a TSDF permitted in Colorado to accept such waste; currently no such permitted facilities are known to be located in Colorado.

The Colorado K903 waste LDR treatment standards apply the federal RCRA characteristics for corrosivity and toxicity, since hydrolysate may exhibit these characteristics. To be defined as K903 the hydrolysate must have a sustained and stable pH > 10. In the Colorado final rule documentation, K901 was amended to include the following language:

> Up to the point in the PCAPP process where the acidic hydrolysate is manipulated to a sustained and stable pH > 10 to ensure destruction of sulfonium ions and TDG-mustard aggregates, prior to transfer from reactor.[5]

[4] Ibid.

[5] 6 CCR 1007-3, Section 268.40.

The current PCAPP RCRA Research and Development and Demonstration (RD&D) Permit Waste Analysis Plan, 2008, Table C-9, indicates that the hydrolysate in the hydrolysate tanks would have a pH between 10 and 12. Therefore, K903 hydrolysate from the reactor that meets the RCRA corrosivity characteristic of pH > 12.5 would have to be treated to meet Colorado LDR treatment standards, including no longer demonstrating the corrosivity characteristic or containing any regulated hazardous constituents above the limits listed in the rule.

However, if the hydrolysate is shipped to an out-of-state TSDF, the waste hydrolysate would have to be treated to meet the LDRs for characteristic wastes as adopted by the state in which the TSDF is located—for example, it no longer exhibits a RCRA characteristic (e.g., corrosivity) *and* meets any additional treatment requirements for known underlying hazardous constituents (e.g., solvents)—prior to final disposal.

Finding 4-1. To meet the definition of a K903 hydrolysate, a nonacute toxic listed waste, the hydrolysate must have a sustained and stable pH > 10 prior to transfer from the reactor for offsite shipment (6 CCR 1007-3, Section 268.40). Otherwise, if the pH ≤ 10, the waste would be defined as K901, an acutely toxic waste, which would impose additional management requirements within Colorado.

Recommendation 4-1. To minimize management requirements while the waste is located within Colorado, PCAPP should ensure the hydrolysate meets the definition of K903.

Structure of RCRA Permit at PCAPP

PCAPP is currently operating under a RCRA RD&D permit because the technologies used in the hydrolysis treatment, coupled with biotreatment of the resulting hydrolysate, are a first-of-a-kind application. An RD&D permit has terms and conditions that will assure protection of human health and the environment, and it limits the treatment to those types and quantities of hazardous waste that the CDPHE deems necessary for purposes of determining the efficacy and performance capabilities of the technology or process and the effects of such technology or process on human health and the environment. RD&D permits can provide for construction of such facilities as necessary and for operation of the facility for not longer than 1 year. The permit can be renewed three times, but each renewal period cannot be for more than 1 year.[6]

Once the performance of PCAPP has been demonstrated to the satisfaction of ACWA, PCAPP operators, and CDPHE, PCAPP must file for a full operating RCRA permit, commonly referred to as a RCRA Part B permit. PCAPP intends to submit the Part B permit application in February 2015.[7]

[6] See 6 CCR 1007-3, §100.25.

[7] Ron Entz, environmental permitting engineer, PCAPP, presentation to the committee on July 30, 2014.

Figure 4-1 shows the additional documentation that is required as PCAPP transitions from the RD&D permit to the full RCRA permit.[8]

Offsite transport and treatment of the hydrolysate by an out-of-state TSDF, although allowed by RCRA regulations as administered by CDPHE, is not described or contemplated in the current RD&D permit nor is it addressed in the current full RCRA Part B permit application documentation.[9]

As noted by the National Research Council (NRC, 2008), PCAPP has acknowledged that a modification to eliminate the use of the biotreatment unit and to add a loading facility for offsite hydrolysate shipments would probably be a Class 2 or Class 3 modification. That report contained the following finding and recommendation:

> **Finding 4-7.** The applications for modifications of the research, development, and demonstration permits for both BGCAPP and PCAPP proposed that a number of specific secondary wastes be shipped offsite for treatment or ultimate disposal.
>
> **Recommendation 4-2.** The Program Manager for Assembled Chemical Weapons Alternatives should continue to pursue the acceptance of the planned offsite shipment and disposal of secondary waste through permit modifications and stakeholder involvement.

During committee discussions at the July 2014 meeting in Pueblo, the CDPHE indicated that if offsite transport of hydrolysate to a TSDF for treatment is necessary, either under the RD&D permit or under the full RCRA operating permit, the process would require a RCRA Class 2 permit modification.[10,11]

For a Class 2 permit modification,[12] the RCRA procedures for public participation require the permittee to send a notice of the modification request to all persons on the facility mailing list maintained by the director and to the appropriate units of federal, state, and local government as specified in the regulations and to publish this notice in a major local newspaper with general circulation. The notice must include the announcement of a 60-day comment period; the date, time, and place of a public meeting to be held in accordance with the regulations; the location where copies of the modification request and any supporting documents can be viewed and copied; and facility contact information.[13] The typical time required for a Class 2 permit modification would be 3 to 6 months, whereas the typical time for a Class 3 modification would be 6 months to a year or more.

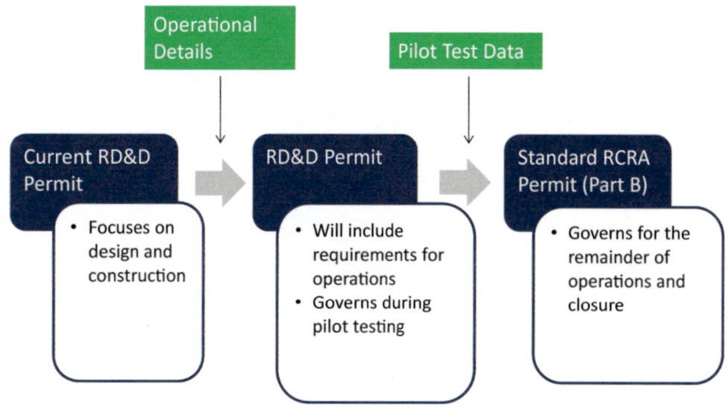

FIGURE 4-1 Transitioning of RCRA permit. SOURCE: Ron Entz, environmental permitting engineer, PCAPP, "RCRA Permit Structure and Potential Modifications for Offsite Shipment of Hydrolysate," presentation to the committee on July 30, 2014.

The time required to implement a Class 2 modification to a RCRA permit may vary depending on regulator and public acceptance. More time would likely be needed for Class 2 or 3 if there is any controversy associated with the modification being proposed.[14]

CDPHE may also issue a Temporary Authorization[15] having a term of up to 180 days. In most cases, Temporary Authorizations, if issued by the regulator, allow site preparation, construction, and potentially similar activities as a permit modification is being considered. Temporary Authorizations to begin construction may be especially important if offsite treatment of hydrolysate becomes necessary at PCAPP, because offsite treatment would likely require construction of a truck loading station and piping and other modifications.[16]

It is most likely that any issues with the on-site treatment of the hydrolysate would arise during systemization or while the facility is still operating under the RD&D permit. The RD&D permit affords the facility more flexibility within the regulatory process than would the full RCRA permit

[8] Ibid.

[9] Ibid.

[10] Doug Knappe, CDPHE PCAPP, permitting unit leader, presentation to the committee on July 30, 2014.

[11] CDHPE, in the supporting documentation for adopting the addition of the K903 hydrolysate listing, specifically acknowledged that "in the event the mustard agent hydrolysate cannot be successfully treated at PCAPP, the waste may be shipped offsite to another permitted hazardous waste treatment, storage or disposal facility that may also manage the waste in accordance with the new hydrolysate listing." (Solid and Hazardous Waste Commission/Hazardous Materials and Waste Management Division, 6 CCR 1007-3, Hazardous Waste, Addition of K903 (Hydrolysate) Listing, Adopted by the Solid and Hazardous Waste Commission on November 20, 2012.)

[12] 6 CCR 1007-3, §100.63(e).

[13] See 6 CCR 1007-3, §100.63.

[14] While the permittee identifies the modification class in the modification application, the CDPHE makes the final decision on the appropriate class of the modification. The regulations for a Class 2 modification provide for a 60-day comment period and the CDPHE has no more than 90 days after the receipt of the modification request to approve or deny the modification, with or without changes. CDPHE could also determine that the modification must following procedures for a Class 3 modification, instead of a Class 2. In addition, the CDPHE can notify the applicant of a 30-day extension for a decision; up to 120 days.

[15] 6 CCR 1007-3 §100.63(e)

[16] Ron Entz, environmental permitting engineer, PCAPP, "RCRA Permit Structure and Potential Modifications for Off-site Shipment of Hydrolysate," presentation to the committee on July 30, 2014.

to try to overcome internal technical obstacles in order to continue the onsite treatment of the hydrolysate, including operational and infrastructure changes to mitigate poor or nonperformance. However, if it appears that these obstacles might not be overcome for a significant time or at all and if it appears that offsite transport of the hydrolysate may be necessary while the plant is under the RD&D permit, PCAPP still would need to wait up to 6 months or more for approval of a Class 2 permit modification to allow offsite transport of hydrolysate. Munitions processing will be delayed during this period. While a temporary authorization may be issued by CDPHE for such a modification, normally this would only allow limited activities to prepare for offsite shipments until the final permit modification is approved (e.g., planning or preconstruction of truck loading facilities).

Finding 4-2. If it appears that obstacles for onsite treatment cannot be overcome and if it appears that offsite transport of the hydrolysate will be necessary, munitions processing would be delayed for 6 months or more based on the regulatory approval process for a Class 2 permit modification.

Recommendation 4-2. PCAPP should process a permit modification for the RCRA Research and Development and Demonstration (RD&D) permit that would allow for the offsite transport of hydrolysate as a backup plan. The modification application should contemplate a temporary authorization for site preparation, preconstruction, and similar activities while PCAPP is operating under the RD&D permit.

Recommendation 4-3. PCAPP should include provisions for the capability for offsite transport of hydrolysate in the PCAPP Part B permit application that PCAPP is preparing.

Any changes to the PCAPP facility, particularly those that require infrastructure changes, would require a RCRA permit modification. For instance, based on findings in Chapter 7, installation of additional onsite hydrolysate storage capacity may be considered to allow more time for remedying poor or nonperformance of components of the biotreatment area. In this example, with additional hydrolysate storage capacity, munitions destruction and agent hydrolysis could continue. However, with specific exceptions, the waste cannot be stored for longer than 1 year under the LDR regulations.

The committee notes, however, that while adding storage capacity would buy more time, if fixes to downstream processes are unsuccessful, the time and resources spent approving and installing additional storage capacity, to include permitting and other documentation, would be wasted. In this instance, PCAPP would still need to consider implementing the offsite option to ensure continued destruction of the stockpile.

Recommendation 4-4. PCAPP should prepare and process any modifications to the RD&D permit or the Part B permit application that would be needed to allow implementation of additional contingency options that may be determined to be prudent solutions to poor or nonperformance of BTA components, including additional hydrolysate storage.

The NRC (2008) report already noted that

> On the basis of discussions with state regulators, Mitretek concluded that if offsite shipment of hydrolysate is adopted, neither BGCAPP nor PCAPP would be allowed to begin operations until an appropriate TSDF had been selected and a contract for receipt of the waste was in place (Bizzigotti et al., 2006).

At this time, even if the permit modification is approved for offsite shipment of hydrolysate, PCAPP has not identified an offsite TSDF that would be available to accept shipments of PCAPP hydrolysate. Identifying and contracting with an appropriate TSDF will take time. The TSDF must have characterization data on the waste to demonstrate that it can accept the waste for treatment under its current RCRA permit. If the PCAPP hydrolysate falls outside the TSDF's normal waste acceptance criteria, that TSDF may itself have to process a RCRA permit modification. In addition, it may take weeks or even months for PCAPP to establish a contracting arrangement with the offsite TSDF, as federal contracting system requirements must be satisfied.

Finding 4-3. Identifying and contracting with an appropriate offsite TSDF that can accept the hydrolysate under its RCRA permit will require time, perhaps weeks or even months, which would prevent the expeditious implementation of offsite shipments and delay the destruction mission.

Recommendation 4-5. PCAPP should proceed as soon as possible to identify at least one acceptable offsite TSDF that is capable of accepting PCAPP hydrolysate and should establish mechanisms for contracting with appropriate TSDFs.

Another potential regulatory issue deals with the reuse of the water recovered after the dewatering of the biotreatment sludge in the brine reduction system (BRS). The RD&D permit currently requires that the BRS be operated so as to ensure that the effluent is of sufficient quality to allow it to be recycled to other portions of the PCAPP and be used as a substitute for well water.

The permit requires that samples be analyzed for either (1) the constituents listed in Appendix VIII of 6 CCR 1007-3, Part 261 or (2) an alternative list of constituents approved by CDPHE. The results should be compared with the limits for constituents found in both the Colorado Primary Drinking Water Maximum Contaminant Levels (MCLs) and the Maximum Concentrations of Contaminants for Toxicity Characteristics.[17]

[17] 6 CCR 1007-3, §261.24.

The process, as outlined in the permit, would allow automatic approval for recycling if the effluent meets the MCLs for those constituents that are also listed in Appendix VIII of Part 261. However, if the effluent cannot meet the MCLs, the results are to be compared with the Maximum Concentrations of Contaminants for RCRA Toxicity Characteristics. If they are below those regulatory levels, the results are to be transmitted to the CDPHE for confirmation before the effluent is considered to not be a hazardous or solid waste and to therefore be recyclable through the plant.

If the effluent does not meet the MCLs and does not meet the Maximum Concentrations of Contaminants for RCRA Toxicity Characteristics, PCAPP must stop any transfer of BRS effluent to downstream systems and must provide CDPHE with a corrective action plan before any transfer can resume. This would result in the facility needing to extract significantly more well water than planned and could delay destruction operations.

During presentations to the committee on July 29 and 30, 2014, PCAPP officials expressed some concern regarding whether the recovered water would be able to meet the primary drinking water MCLs and noted that PCAPP was considering initiating a delisting petition for the BRS effluent.[18]

As stated above, hydrolysate is a listed waste (K903). Therefore, under the derived-from rule, residuals generated by the hydrolysate treatment process would carry the K903 hazardous waste code and would have to be disposed of as a hazardous waste. It should be noted that materials that are reclaimed from solid wastes and that are used beneficially are not solid wastes and hence are not hazardous wastes unless the reclaimed materials are burned for energy recovery or used in a manner constituting disposal.[19] In addition, the CDPHE may grant requests for a variance from classification as a solid waste for those materials that are reclaimed and then reused as feedstock within the original production process in which the materials were generated if the reclamation operation is an essential part of the production process.[20] Therefore, because the water is intended for reuse within the system and never leaves the PCAPP processing areas, it could be argued that it is not a waste material. The only issue with the recovered water should be whether its quality is sufficient to support PCAPP engineering processes.

Finding 4-4. Because the water exiting the BRS is intended for reuse within the system and never leaves the PCAPP processing areas, it should not be considered a waste material. The only issue with the recovered water should be whether its quality is sufficient to support PCAPP engineering processes.

[18] Presentation by Paul Usinowicz, Bechtel technical advisor, and Yakup Nurdogan, senior wastewater treatment specialist, July 29-30, 2014, and presentation by Don Guzetti, start-up field supervisor, PCAPP, and Dave deLesdernier, Battelle, July 29-30, 2014.

[19] 6 CCR 1007-3, §261.3(c)(2)(i).

[20] 6 CCR 1007-3, §260.32.

Recommendation 4-6. PCAPP should confirm with the Colorado Department of Public Health and Environment that the effluent recovered from the brine reduction system does not fall under Colorado hazardous listing for hydrolysate (K903); should process any delisting petitions, variance requests, or permit modifications necessary to accomplish this; and should turn its attention to ensuring the recovered water meets technical requirements for reuse within PCAPP.

NEPA REQUIREMENTS

As indicated in NRC (2008), under the National Environmental Policy Act (NEPA), PCAPP prepared and issued an environmental impact statement (EIS) covering the construction and operation of the chemical agent treatment facility. Neither that EIS nor the corresponding records of decision address offsite shipment of hydrolysate. Under NEPA regulations, if the new proposed action is not adequately covered in an existing EIS or environmental assessment, the site would have to prepare an environmental assessment, which would result in either a finding of "no significant impact" or a requirement to prepare supplemental NEPA documentation.

In *The Sierra Club, et al.,* v *Dr. Robert M. Gates, Secretary of Defense, et al.*, in the U.S. District Court of the Southern District of Indiana, Terre Haute Division, the plaintiffs sought to stop the government from continuing shipments of the product of the hydrolysis of the nerve agent VX from the Newport Chemical Weapons Depot to Veolia's incineration facility in Port Arthur, Texas. In that case, the Court ruled that the government published its notice of the Record of Decision regarding the Final Programmatic Environmental Impact Statement (FEIS) for its Chemical Stockpile Disposal Program. That FEIS indicated that site-specific NEPA review, which would include the preparation of an EIS or an Environmental Assessment (EA), would be conducted for each of the chemical storage installations.

The Court found that the Army's 1998 FEIS for the pilot test of its neutralization plan for the VX at the Newport Chemical Agent Disposal Facility (NECDF) in Indiana evaluated two alternatives: no action and the proposed process. Partly in response to certain NRC studies and partly in response to the terrorist attacks of September 11, 2001, in July 2002, the Army published *Final Environmental Assessment* (FEA 2002) regarding the VX destruction process at NECDF. In that FEA, the Army compared a no-action alternative with the disposal of the hydrolysate at an offsite TSDF. The FEA made no findings with respect to a specific TSDF, suggesting that such analysis would be performed later, after a TSDF had been chosen. The Army issued *Final Finding of No Significant Impact* (FONSI 2002) in October of that year.

In 2003, the Army issued a transportation analysis that analyzed the environmental impacts for two routes from Newport, Indiana, to Deepwater, New Jersey, to reach the DuPont environmental treatment facility. In 2005, the Army issued a record of environmental consideration (REC), announcing that it would start neutralization of the VX at NECDF and that the hydrolysate generated from the process would be stored onsite until a suitable arrangement could be made for its treatment at a TSDF; the REC indicated its decision was covered by FEA 2002 and FONSI 2002.

In 2006, after extensive analysis and discussions with the receiving TSDF, the EPA approved the offsite treatment option. In April 2007, the Army issued another REC for the proposed action to ship caustic wastewater derived from the destruction of VX at NECDF to a permitted commercial TSDF and found that the proposed action qualified for a categorical exclusion for routine management of hazardous materials/hazardous waste operations.[21] The Army relied on previous NEPA documentation to support the April 2007 REC, including previous FEIS and FEA documents. In June 2007, the Army issued another REC in response to a letter from plaintiffs concerning the safety and environmental impacts of the proposed action and used the same rationale for not performing an EIS or EA—that the hydrolysate is classified as hazardous waste and comes under the Army's categorical exclusion for regular treatment and handling of hazardous waste from its facilities.

The Court ruled that it must defer to an agency's factual findings when deciding whether the environmental impacts of its actions are significant and may not substitute its own judgment for that of the agency. Under such a review, the Court agreed with the government that it did not need to supplement its FEIS 1998 or the FEA 2002 or provide an additional comment period when the alternative option was proposed. The Court found that the administrative record reflected that the original NEPA documents considered onsite treatment of the hydrolysate versus shipment of the hydrolysate offsite to a permitted TSDF. The Court also found that the secondary decision to switch to another permitted TSDF[22] does not need a supplemental EIS or EA because the government had taken the necessary "hard look"[23] at the nature of the hydrolysate and correctly determined that it was hazardous waste.

Any permit modification to conduct the offsite shipment of hydrolysate at PCAPP would require consideration under NEPA. Unlike NECDF, the NEPA documentation provided by PCAPP did not include an alternative action for offsite shipment of hydrolysate. PCAPP may find that it needs to take a hard look at the nature of the hydrolysate and the action of shipping it to an offsite TSDF. Then it should determine if such an action would be covered by its current NEPA documentation or if it should come under the categorical exclusion or if supplemental NEPA documentation would be necessary.

Finding 4-5. If it is determined that the BTA cannot treat the hydrolysate onsite and it becomes necessary to consider shipping hydrolysate offsite, a corresponding NEPA action may be necessary, potentially delaying shipment of hydrolysate and, once storage capacity is reached, the destruction operations at PCAPP would need to halt.

Recommendation 4-7. PCAPP should immediately initiate a study to determine whether it will need to amend its current National Environmental Policy Act (NEPA) documentation or if a new action is warranted, including issuing a Final Environmental Assessment, a Final Programmatic Environmental Impact Statement, or establishing the basis for a categorical exclusion. Such a study should be completed while PCAPP is operating under its RCRA RD&D permit. PCAPP should determine the time frame the selected NEPA action would require and mitigate any delay by prefiling if it is determined that a new proposed action under NEPA is required.

PUEBLO COUNTY CERTIFICATE OF DESIGNATION

PCAPP was required to obtain a certificate of designation from the Pueblo County Board of County Commissioners before beginning operations. The county grants such a certificate only after the CDPHE has reviewed and recommended approval of the facility.[24] The county also reviews written statements and supporting documentation on whether the proposed facility poses a risk to the health and/or the safety of the public and/or the environment; the density of the population neighboring the facility and the routes to and from the facility; the risk of accidents; and compliance with zoning requirements and the impact on neighboring activities—for example, agricultural activities, infrastructure, and community support services. It also considers the facility owner's financial responsibility and prior performance records.

As for the CDPHE permitting processes, the certificate holder must notify the county of any modification or changes in operations, ownership, or design for a hazardous waste processor that involve matters that are the subject of or contained in the Certificate of Designation. Within 10 days of the receipt of a modification notice, the county must notify the certificate holder of the classification of the modification: Class A, which requires no additional information or input from the certificate holder; Class B, which needs additional information or input and may require either an informal modification that needs only additional informa-

[21] A categorical exclusion is defined as actions that normally do not require an EA or EIS, and the Army has determined that they do not individually or cumulatively have a substantial effect on the human environment. From Appendix B of 32 CFR Part 651 (AR 100-2), *Environmental Analysis of Army Actions*.

[22] The Army switched from shipping to the DuPont treatment facility in Deepwater, New Jersey, to shipping to the Veolia incinerator facility in Port Arthur, Texas.

[23] Courts consistently have held that, at a minimum, NEPA imposes a duty on federal agencies to take a "hard look at environmental consequences" (Natural Resources Defense Council v Morton, 458 F.2d 827, 838 D.C. Cir., 1972).

[24] See CRS 30-20-100 and Pueblo County Code Section 17.176.090.

tion, explanation, or discussions or a formal modification, which requires a more detailed review process resulting in an amendment to the Certificate of Designation, including a 20-day public notice and comment period and public review and, possibly, a hearing following the public comment period and/or amendment to the Certificate of Designation; or Class C modification, which is for a modification of ownership, design, or operations so substantial that an amendment process is warranted. There is no provision for a Temporary Authorization, as is allowed under RCRA permitting, under the Pueblo County Code.

Finding 4-6. If a modification to the current RCRA RD&D permit becomes necessary in order to ship the hydrolysate offsite, a corresponding application for modification to the Certificate of Designation would have to be made to the Pueblo County Board of Commissioners.

Recommendation 4-8. PCAPP should initiate an action with the Pueblo County Board of Commissioners to prepare such a modification application while it is operating under its RCRA RD&D permit.

ORGANISATION FOR THE PROHIBITION OF CHEMICAL WEAPONS

The United States is a signatory to the Chemical Weapons Convention (CWC), which is overseen by the Organisation for the Prohibition of Chemical Weapons (OPCW). The treaty is administered through the U.S. Department of State. As a U.S. stockpile destruction site, PCAPP is subject to onsite monitoring of the destruction process by OPCW representatives. Not only is the mustard itself subject to the CWC, but also subject to it are what are known as Schedule 2 compounds, which include chemicals that may be used to manufacture chemical agents, or chemical agent precursors. Many of these Schedule 2 chemicals are also used in industrial processes to produce other products. TDG, which can be used in the manufacturing process for mustard, is also a common hydrolysis product contained in mustard hydrolysates and, therefore, is a Schedule 2 chemical under the CWC.

The primary purpose of the biotreatment system at PCAPP, to which mustard hydrolysate is fed, is to mineralize the organic compounds present in the hydrolysate, including TDG. TDG resulting from the hydrolysis process at PCAPP is expected to be present in mustard hydrolysate at a concentration of around 56,000 mg/L, or parts per million (ppm). The hydrolysate fed into the immobilized cell bioreactors is to be diluted for an effective concentration of approximately 7,000 ppm TDG. The current PCAPP RD&D RCRA permit requires the treated hydrolysate to be sampled in the immobilized cell bioreactor (ICB) effluent tanks and then analyzed for TDG and total organic carbon (TOC) to determine the effectiveness of the biodegradation process. The performance goals for the biodegradation process are established in the RCRA permit as achieving a performance goal of greater than 95 percent and a target of at least 99 percent removal of the agent hydrolysate product, TDG, and an average of at least 90 percent removal of TOC.[25] In addition, according to PCAPP presentations, the goal of the treatment process is to achieve ≤0.1 wt% TDG. At present, however, it is unclear whether the OPCW agrees with any of the above stated treatment goals for TDG treatment.[26]

During the pilot test of the first-of-a-kind application of these biotreatment technologies beginning in September 2014 through July 2015 (see Chapter 6), PCAPP intends to feed a hydrolysate surrogate into the biotreatment area processes at PCAPP (PCAPP, 2014). PCAPP will procure TDG from a manufacturer in Germany to conduct these tests.

The industrial grade TDG to be used for Biotreatment Area Risk Reduction and Mitigation, as a Schedule 2 chemical, is subject to the requirements of the CWC. Under the *CWC Industry Inspection Preparation Handbook*, any facility that produces, processes, or consumes in excess of 10 metric tons of any Schedule 2 chemical is subject to inspection. Accordingly, it will be subject to oversight by the OPCW.[27] Under OPCW export requirements for TDG, a mixture containing 20 percent of TDG does not require a license for export to a state party to the CWC but is prohibited from being exported to a state not party to the CWC. A mixture containing 35 percent requires a license for export to certain state parties. A mixture containing less than 7 percent does not require a license for export to any destination, including a state not party to the CWC.[28]

The PCAPP plan is to treat the hydrolysate surrogate in the biotreatment process as if it was actual mustard hydrolysate. Following a number of pretests and demonstrations, actual testing with surrogate will begin in March 2015. PCAPP will then transition from treatment of hydrolysate surrogate to actual mustard hydrolysis beginning in mid-September 2015.[29]

Finding 4-7. The RCRA permit performance goal for the biotreatment process is >95 percent and a target of at least 99 percent of removal of agent hydrolysate product, TDG as

---

[25] PCAPP RCRA RD&D Stage III, Class 3, Permit Modification Request, Revision 0, November 2006, Appendix D, Waste Analysis Plan, C-2a(3)(d), p. C-11.

[26] Paul Usinowicz, Bechtel technical advisor, and Yakup Nurdogan, senior wastewater treatment specialist, presentation to the committee on July 30, 2014.

[27] Under the *CWC Industry Inspection Preparation Handbook*, any facility that produces, processes, or consumes in excess of 10 metric tons of any Schedule 2 chemical is subject to inspection.

[28] Department of Commerce, Bureau of Industry and Security, Chemical Weapon Convention: Import and Export Requirements, http://www.cwc.gov/outreach_industry_publications_cwc007.html#F_6_.

[29] Don Guzzetti, start-up field supervisor, PCAPP, and Dave deLesdernier, Battelle, presentation to the committee on July 30, 2014.

measured in the ICB effluent tanks.[30] The PCAPP Pilot Test Demonstration Plan proposes a goal of ≤0.1 wt% for TDG in the effluent.[31] The PCAPP presentation indicated criteria of >95 percent TDG removal efficiency at design flow rate and ≥98 percent TDG removal efficiency at 50-90 percent design capacity and ≤0.1 wt% TDG when the ICB unit is being air scoured or in recovery.[32]

Finding 4-8. At present, it is unclear whether the OPCW agrees with any of the above treatment goals for the TDG present in the hydrolysate.

Finding 4-9. The testing process with hydrolysate surrogate will result in waste streams similar to those that will be produced during the actual mustard treatment campaigns at PCAPP and that will feed into the ICBs, the BRS, and the water recovery system, ultimately producing filter cake.

Recommendation 4-9. ACWA should work with the CDPHE and, if necessary, OPCW to demonstrate the efficacy of the treatment goals for thiodiglycol during the hydrolysate surrogate testing portion of the overall biotreatment area risk reduction and mitigation process. PCAPP should also work with regulators to establish a final treatment standard or goal in the RD&D permit and the final Part B permit.

REFERENCES

Bizzigotti et al. 2006. *Analysis of Offsite Treatment of Hydrolysates from Chemical Agent Destruction Pilot Plants.* Mitretek report MTR 2006-22, October. Falls Church, Va.: Mitretek Systems.

NRC (National Research Council). 2008. *Review of Secondary Waste Disposal Planning for the Blue Grass and Pueblo Chemical Agent Destruction Plants.* Washington, D.C.: The National Academies Press.

PCAPP (Pueblo Chemical Agent Destruction Pilot Plant). 2014. *White Paper Bio-Treatment Area Risk Reduction and Mitigation.* 24852-30H-BTA-V0001. Rev. 000. April.

[30] PCAPP RCRA RD&D Stage III, Class 3, Permit Modification Request, Revision 0, November 2006, Appendix D, Waste Analysis Plan, C-2a(3)(d), p. C-11.

[31] PCAPP Pilot Test Demonstration Plan, CDRL H002, 24852-GPP-GYPM-00006, Revision 000, August 2013, Table 2-1, p. 2-5.

[32] Presentation by Paul Usinowicz, Bechtel technical advisor, and Yakup Nurdogan, senior wastewater treatment specialist, July 29-30, 2014, and presentation by Don Guzzetti, start-up field supervisor, PCAPP, and Dave deLesdernier, Battelle, July 29-30, 2014.

5

Transportation of Chemical Materials

This chapter begins with a description of the federal regulations relevant to the definition and reporting of heavy truck accidents (crashes) and hazardous material incidents. Some historical offsite transportation data for hydrolysate from chemical agent destruction sites is then presented. The risks of transporting hydrolysate, addressed next, entail the following: (1) those due to a heavy truck crash independent of the cargo, (2) those due to a hazardous material cargo, and (3) those unique to hydrolysate.

REGULATIONS OF THE U.S. DEPARTMENT OF TRANSPORTATION

The U.S. Department of Transportation (DOT) is the federal government agency with the primary responsibility for policies and programs to protect and enhance the safety, adequacy, and efficiency of the transportation system and services. DOT consists of 11 individual operating administrations, including the Federal Aviation Administration (FAA), the Federal Motor Carrier Safety Administration (FMCSA), the Federal Railroad Administration (FRA), and the Pipeline and Hazardous Materials Safety Administration (PHMSA). The PHMSA is responsible for hazardous materials regulations, including classification of hazardous materials into one of nine classifications, the associated placarding of vehicles, packaging requirements, and so on (49 CFR 171-180).

Hydrolysate shipments to date have been by truck, and the emphasis in this chapter is therefore on truck transport across public roads and highways. The shipment of hydrolysate by rail is potentially an option and will have similar risks to the public—that is, the potential for direct physical impact and the potential for release of the hazmat cargo. The purpose of this overview of DOT regulations is to introduce the concepts of a reportable crash, a reportable incident, and the first step of hazard classification that dictates subsequent regulations to be followed.

Of particular relevance to this report is that the FMCSA is responsible for maintaining a database to provide information on serious crashes of trucks (and buses). A crash is reported to FMCSA ("DOT-reportable") if it involves the following (49 CFR 390.05):

- Any truck having a gross vehicle rating of more than 10,000 pounds used on public highways or any vehicle displaying a hazardous material (hazmat) placard.

and

- That vehicle is involved in a crash while operating on a roadway customarily open to the public, which results in any of the following:
  —A fatality as a result of the crash;
  —An injury requiring medical treatment away from the crash scene; or
  —A tow-away of any motor vehicle disabled as a result of the crash.

Further, a reportable hazmat incident is defined (49 CFR 171.15) and reported (49 CFR 171.16) separately to PHMSA if, as a result of a hazmat,

- A person is killed or receives an injury requiring admittance to a hospital;
- The general public is evacuated for 1 hour or more;
- A major transportation artery is closed for 1 hour or more;
- An unintentional release of a hazmat or the discharge of any quantity of hazardous waste; or
- A specification cargo tank of 1,000 gallons or more containing hazmat either suffers damage to the lading retention system or requires repair to a system intended to protect the lading retention system even if no release occurs.

The above characteristics are important when the transportation of chemical munition materials is considered.

Other parts of the definition—for example, those involving radioactive materials—do not apply to this report.

The definition of "accident" in 49 CFR 390.5 uses "occurrence" instead of "crash." Further, "DOT-reportable" usually includes "accident." However, many DOT documents use "crash" in lieu of "accident" to clearly indicate that physical forces are involved (DOT, 2013). Note that an incident may or may not involve a crash but does involve a hazmat release. In this chapter the committee uses "accident" in the phrase "DOT-reportable accident" and "crash" elsewhere unless it is quoting.

Hydrolysate shipments to date in the chemical demilitarization program (Table 5-1) have been considered Class 8 "corrosive" hazmat. Class 8 hazmat is defined in 49 CFR 173.136 as a liquid or solid that causes (1) full thickness destruction of human skin within a specified period of time or (2) a specified corrosion rate of steel or aluminum. The rate of destruction defines the Packing Group as I, II, or III. The class and packing group then dictate a number of important DOT requirements (e.g., equipment selection and inspection procedures); however, a detailed discussion of DOT hazmat regulations is beyond the scope of this chapter. The DOT regulations do not explicitly equate corrosive with a pH value, but the U.S. Environmental Protection Agency (EPA) defines corrosive hazardous waste as (1) a liquid with pH <2 or >12.5 or (2) a liquid that corrodes steel at a rate of >0.250 in./yr at a test temperature of 55°C (130°F). The EPA and DOT definitions are frequently confused.

HISTORICAL TRANSPORTATION OF CHEMICAL MUNITION MATERIALS

The anticipated composition of the Pueblo Chemical Agent Destruction Pilot Plant (PCAPP) hydrolysate is compared in Table 5-1 to hydrolysate shipped from Aberdeen Proving Ground (APG) and the Newport Indiana Chemical Depot (NECD) as well as the Explosive Destruction System neutralent from destruction of bomblets at Rocky Mountain Arsenal (RMA) in 2001. Note that the data shown in Table 5-1 for PCAPP hydrolysate characteristics were provided by PCAPP at the committee meetings in July 2014. These parameters may differ from those presented in Table 2-1, which were derived from the 2003 Guelta and Fazekas-Carey study and from the 2006 Waste Analysis Plan filed by PCAPP as part of its Resource Conservation and Recovery Act (RCRA) RD&D permit, which is currently under revision. The differences, however, are not significant for the purpose of comparing the risk of hydrolysate shipments.

Historical offsite transportation data for sarin, distilled mustard agent, and VX nerve agent hydrolysate from Blue Grass Army Depot (BGAD), APG, and NECD as well as neutralent from RMA bomblet destruction are presented in Table 5-2. All shipments in the table were by truck.

The above data show that more than 500,000 shipment miles have accumulated with materials similar to the hydrolysate that will be generated at PCAPP without a leak or even a fender-bender crash.

Finding 5-1. The historical shipment mileage data are dominated by the shipment of hydrolysate from NECD to the Veolia TSDF. The NECD shipments, as well as those from APG, BGAD (Operation Swift Solution), and RMA were free from a minor fender bender or leaks of hydrolysate or similar fluids.

Finding 5-2. Offsite shipments of hydrolysate and similar chemical munition materials from APG, NECDF, BGAD, and RMA have been safe.

IDENTIFICATION OF HYDROLYSATE TRANSPORTATION RISKS

Risk is the combination of likelihood and the consequence of a specified hazard being realized. Likelihood is usually expressed in transportation risk analyses as crashes/mile, crashes/trip, etc., and may be modified by conditional probabilities such as the probability of a fire, given a crash, and the probability of fire causing failure of the hazmat container, given a fire occurs. The associated consequence could be injuries or fatalities from a hazmat release due to a fire. Risk analysis methodologies differ in (1) the scope of the analyses, e.g., selection of the various options available such as mode or container, and (2) the level of detail needed to accomplish the purpose of the analysis, e.g., hazmat release as an appropriate consequence or whether dispersion of the released hazmat and potential exposure to the public needs to be considered also, including the appropriate conditional probabilities. All risk analyses aggregate risk-producing components to some extent depending on the data availability and the purpose of the analysis. Therefore, the risk analysis may be qualitative, semiquantitative, or fully quantitative. The purpose of this chapter is not to completely specify the level of detail needed in a PCAPP quantitative transportation risk analysis (QTRA), although some general requirements are identified at the end of this chapter. Rather, the purpose is to identify how the transportation risk changes with the type of cargo. Some quantification is provided to help evaluate the risk associated with the cargo type.

The risks are identified in this section as (1) fatalities and/or injuries due to the physical impact of a heavy truck with a person independent of the cargo, (2) fatalities, injuries, and/or economic consequences due to the release of the hazmat cargo, and (3) risks unique to a cargo of hydrolysate in addition to (1) and (2). As stated earlier, the shipment of hydrolysate by rail is potentially an option, but for the purposes of this chapter, the focus is on truck shipment across public roads and highways. Risk identification will be similar for truck and rail.

TABLE 5-1 Comparison of Chemical Agent Liquid Treatment Content

Parameter	PCAPP Anticipated Hydrolysate[a]	APG HD Hydrolysate[b]	NECD VX Hydrolysate[c]	EDS Neutralent (GB Bomblets at RMA)[d]
Primary active ingredient	Hot water and NaOH	Hot water and NaOH	Water and NaOH	Monoethanolamine
Water (wt%)	88	88-97	71-91.7	51.7-56.2
Approximate pH	10-13[e]	12.4	12.5-14	12
Thiodiglycol (TDG) (ppm)	56,500	52,250		NA
Isopropyl methylphosphonate (IMPA) (ppm)				3,400-5,000
Diisopropyl methylphosphonate (DIMP) (µg/L)				18,000-27,400
Sodium 2-(diisopropylamino) ethylthiolate (%)			<11	
Sodium ethylmethyl phosphonate (%)			<10	
Sodium methyl phosphonate (%)			<2	
Diisopropylamine (%)			<4	
1,4-Dithiane (ppm)	1,294	1,371		
1,4-Oxathiane (ppm)	512	734		
1,2-Dichloroethane (ppm)	181	181		
Total organic carbon	31,600 ppm	27,875 mg/L	<12%	
Total suspended solids	8,156 ppm	8,676 mg/L	<1.0%	
Sodium chloride (NaCl) (ppm)	54,000-61,000			
Benzene	301 ppb	319 ppb		1,300-2,850 µg/L
Chloroform	301 ppb	329 ppb		ND-21.6 µg/L
Dichloromethane (µg/L)				ND-97.1
Toluene	150 ppb	58 ppb		369-810 µg/L
Vinyl chloride (ppm)	11	12		
Ammonia (ppm)			<500	
Arsenic	1,806 ppb	2,297 ppb	<5 ppm	<200 µg/L
Barium (ppm)			<100	
Cadmium		95 ppb	<1 ppm	6.81-10 µg/L
Chromium	1,505 ppb	1,639 ppb	<5 ppm	445-770 µg/L
Copper	6,170 ppb	6,515 ppb	<1 ppm	9,030-18,200 µg/L
Lead	1,354 ppb	1,377 ppb	<5 ppm	63-237 µg/L
Mercury	150 ppb	164 ppb	<0.2 ppm	0.1-1 µg/L
Iron (ppm)	2,031	2,161	<5	
Selenium (ppm)			<1	
Silver (ppm)			<5	
Zinc	3,611 ppb	3,811 ppb	<10 ppm	23,100-38,300 µg/L
Explosives in liquids (µg/L)				<1,000

NOTE: NaOH, sodium hydroxide (caustic); ND, not detected; ppm, parts per million; ppb, parts per billion; EDS, Explosive Destruction System.

[a] PCAPP PFD 24852-RD-M5-B04-B0004 and 24852-RD-M5-B09-B0002.

[b] Aberdeen Chemical Agent Disposal Facility shipment analysis data for shipments between June 14, 2004, and February 9, 2005, provided by Bill Steedman.

[c] December 12, 2006, Waste Characterization Sheet.

[d] Laurence Gottschalk, Director, Recovered Chemical Materiel Directorate, presentation to the committee on July 30, 2014.

[e] Don Guzzetti, start-up field supervisor, PCAPP, "Biotreatment Area Risk Reduction and Mitigation," presentation to the committee on July 29, 2014.

SOURCE: Adapted from data provided by PCAPP on July 30, 2014; subreferences as noted.

TABLE 5-2 Historical Shipment Data

Parameter	Operation Swift Solution GB Hydrolysate (JPEOCBD, 2014)	APG HD Hydrolysate (JPEOCBD, 2014)	NECD VX Hydrolysate (JPEOCBD, 2014)	RMA Bomblet Destruction (RCMD, 2014)
Origin	BGAD	APG	NECDF	RMA
Destination	Veolia TSDF, Port Arthur, Tex.	DuPont TSDF, Deepwater, N.J.	Veolia TSDF, Port Arthur, Tex.	Safety-Kleen TSDF, Deer Park, Tex., or APG, Md.
Number of shipments	2	Approximately 1,450	424	2/1
One-way mileage	1,140	49	1,011	1,032/1,705
Total shipment mileage	2,280	Approximately 69,580	428,664	3,769
DOT label and marking (flash point <200ºF if applicable)	Class 8, Packing Group II, waste corrosive liquid, basic organic, n.o.s., UN3267, RQ (sodium hydroxide)	Class 8, Packing Group II, corrosive liquids, n.o.s. (TDG + 5% NAOH solution + D16), UN1760	Class 8, Packing Group II, waste corrosive liquid, basic organic, n.o.s., UN3267, RQ (sodium hydroxide)	Uncertain
DOT reportable accidents (crashes)	None reported	None reported	None reported	None reported
Incidents	None reported	None reported	None reported	None reported
Nonreportable crashes (fender benders)	None reported	None reported	None reported	None reported

NOTES: n.o.s., not otherwise specified; JPEOCBD, Joint Program Executive Office for Chemical and Biological Defense; GB, sarin; VX, a nerve agent; HD, distilled mustard agent; TSDF, treatment, storage, and disposal facility; RCMD, Recovered Chemical Materiel Directorate.

Risks Due to a Heavy Truck Crash Independent of the Cargo

The likelihood that a large truck would be involved in a serious crash is about 1 in a million miles (1.0×10^{-6}/mi). Given a serious heavy truck crash, the probability of (1) a fatality is about 0.01 (1 percent) and (2) an injury is 0.22 (22 percent) independent of the cargo (DOT, 2013).

Finding 5-3. Injuries and fatalities due to the physical forces involved in a heavy truck crash, independent of the cargo, are a risk.

Hazmat crash rates were found to be about half the rate for heavy trucks in general (Battelle, 2001). The current crash rate (DOT, 2014b) for Tri-State Motor Transit Co. (the parent company is Bed Rock Inc.), a company frequently used to transport hazmat and the carrier for NECDF, is 0.30 in a million miles (0.30×10^{-6}/mi) or one-third the rate for heavy trucks in general. This rate applies to the company's entire fleet, not just hazmat.

Hydrolysate shipments from NECDF and APG were subjected to enhanced safety measures as compared with typical hazmat shipments—that is, safety inspections were carried out once every 2 hours or so. The same safety measures can be expected if hydrolysate is shipped from PCAPP, and a crash rate $<0.30 \times 10^{-6}$/mi can be expected.

Table 5-3 estimates the number of PCAPP heavy truck shipments with onsite treatment of hydrolysate at 176/month. Table 5-3 also estimates 211 monthly shipments if hydrolysate is shipped offsite. This estimate of 211 shipments is probably an upper bound because the number of caustic shipments in the case of offsite hydrolysate treatment was not reduced owing to the committee's lack of data about the extent of the reduction in the amount of caustic needed, given offsite shipment of hydrolysate. The likelihood of an injury or a fatality provided above applies to the 176 shipments as well as the 211 shipments.

Finding 5-4. Offsite hydrolysate transport would increase the number of shipments from about 176 per month to as many as 211 per month.

Additional Risks Posed by a Truck Carrying Hazmat

In addition to the cargo-independent consequences of a large truck crash, the potential consequences of a release of hazmat include injuries, fatalities, and cleanup costs. As noted in a recent report (TRB, 2013), "Hazmat-specific accident rates are usually not available and truck accident rates are often used as a proxy." These rates and conditional release probabilities are not known, in part because private as well as public stakeholders (i.e., the Transportation Security

TABLE 5-3 Estimated PCAPP Truck Shipments (trucks per month)

	With Onsite Hydrolysate Treatment (PMACWA, 2012)		With Offsite Hydrolysate Treatment	
Facility/Material/In or Out	Hazardous	Nonhazardous	Hazardous	Nonhazardous
All but ICB and BRS/in and out	8	119	8	119
Caustic tank	24		24[a]	
ICB/DAP/in		1/4		
ICB/carbon/in		1/2		
ICB/urea/in		1		
ICB/carbon/out	0.5			
ICB/biomass/out		0.10		
ICB/hydrolysate/out			60[b]	
BRS/H_2SO_4/in	1			
BRS/conditioner/in	0.25			
BRS/carbon/in		0.5		
BRS/carbon/out	0.5			
BRS/filter cake/out	21			
Total[c]	55	121	92	119
Total	176		211	

NOTE: ICB, immobilized cell bioreactor; BRS, brine reduction system.
[a] A significant number of caustic shipments could be eliminated.
[b] PMACWA, 2006.
[c] Rounded.

Agency) protect data for several reasons such as competitiveness and need to know.

When cargo-independent fatalities/injuries are compared with hazmat-cargo-related fatalities and injuries, the hazmat-cargo-related effects can frequently be neglected. This was the case in the 2003 transportation risk analysis of PCAPP solid wastes and hydrolysates (ANL, 2003). The committee's finding and recommendations about this approach, originally presented in 2008, are presented later in this chapter.

Table 5-4 shows a summary of highway incidents in 2013 by transport phase (DOT, 2014a). The number of incidents in transit is about a third of those taking place during loading and unloading, but they account for the majority of fatalities and monetary damages.

More than 800,000 highway hazmat shipments occur each day (DOT, 2004), which is about 300 million every year. The number of annual highway incidents (13,873) is very small compared with the number of highway hazmat shipments.

Finding 5-5. Historically, the risk of hazmat release during transportation owing to either a leak due to a crash or a crash-independent leak has been small.

Additional Risks Posed by a Truck Carrying Hydrolysate

If hydrolysate is to be shipped offsite, approximately 1,400 tanker truckloads, each holding about 6,000 gal, would be required, or about two tanker trucks a day, every day, for approximately 2 years (PMACWA, 2006). The hazards due to hydrolysate exposure are considered moderate. In comparison, the hazard for 50 percent sodium hydroxide, a typical Class 8 material, is considered high (Noblis, 2008). Therefore, the discussion above regarding the risks of hazmat shipment deals with the risk of hydrolysate shipment.

Pueblo County requires that the "risk of accidents during the transportation of any wastes to, from, or at the proposed site . . . be considered " (Pueblo County land use regulations at Title 17, Chapter 176, Section 050). Noblis (2008) reported that in a 2007 meeting with stakeholders, the attendees indicated that the executive director of CDPHE "wants an assessment performed of the risk associated with the transport of hydrolysate within the State of Colorado." The committee has no knowledge of the specific basis for the request and believes that the 2003 QTRA and subsequent hydrolysate transportation safety evaluations are publicly available.

TABLE 5-4 Highway Hazmat Incident Summary by Transportation Phase in 2013

Highway Transportation Phase	Incidents	Hospitalized	Not Hospitalized	Fatalities	Damages ($)
In transit	2,921	5	28	10	40,988,245
In transit storage	261	8	0	0	565,698
Loading	3,156	4	20	1	2,505,998
Unloading	7,535	4	70	0	4,277,732
Total	13,873	21	118	11	48,337,673

RELATED PRIOR NRC FINDINGS AND RECOMMENDATIONS

The QTRA prepared for PCAPP, referred to earlier in this chapter (ANL, 2003), qualitatively dismissed the risks of transporting hydrolysate compared with the cargo-independent risks from heavy trucks. The 2008 NRC report said it was important to provide quantitative data to calm the anxiety that could be triggered by the prospect of offsite transportation. The 2008 report contained the following finding and recommendations, applicable to PCAPP:

> **Finding 6-4.** Some members of the public and state regulators are concerned about the health risks of hydrolysate transport and believe there is a need for emergency planning along the route.
>
> **Recommendation 6-3.** The PMACWA should perform a quantitative transportation risk assessment for hydrolysate, including a quantitative assessment of the human health consequences of hydrolysate spills with and without a fire. This assessment needs to be completed to facilitate discussions with the public and regulators about the hydrolysate offsite shipment alternative.
>
> **Recommendation 6-4.** The PMACWA should prepare a prototypical emergency response plan for hydrolysate shipment, including the possibility of a fire or the occurrence of natural disasters such as floods. This plan would be the starting point for setting contractual requirements for the TSDF and the shipper. The prototype needs to be completed to facilitate discussions with the public and regulators about the hydrolysate offsite shipment alternative.
>
> **Finding 6-8.** The experience to date with the offsite shipment and treatment of mustard and nerve agent hydrolysates from the Aberdeen and Newport Chemical Agent Disposal Facilities indicates that offsite transport and disposal of these materials is a safe and technically viable course of action.

Finding 5-7. The findings and recommendations cited above from the previous PCAPP review (NRC, 2008) continue to be relevant.

Recommendation 5-1. Transportation-related recommendations in the previous PCAPP review (NRC, 2008) should be followed.

The Army has sufficient experience transporting hydrolysate from APG and NECD to perform a reasonable QTRA in the near term including selection of the packaging (container). The primary missing piece is the selection of a TSDF for the analysis, and Army procurement regulations may preclude timely identification of the selected TSDF. The committee believes that utilizing several representative TSDFs and performing a QTRA for each will accomplish several objectives, including these: (1) illustrate the sensitivity of the risk to different routes to the different TSDFs, (2) quantify the relative risk contributions of both cargo-independent and hydrolysate release scenarios, (3) quantify the overall risk magnitude of hydrolysate transport to a sufficient degree for regulatory and stakeholder evaluation, and (4) provide input to emergency response planning. As noted above (NRC, 2008), the QTRA should include human health consequences of hydrolysate spills with and without a fire.

REFERENCES

ANL (Argonne National Laboratory). 2003. *Transportation Risk Assessment: Options for the Offsite Shipment and Disposal of Residual Wastes from the Destruction of Chemical Weapons at the Pueblo Chemical Depot.* Prepared for the Program Manager, Assembled Chemical Weapons Assessment. APG, Md. April.

Battelle. 2001. *Comparative Risks of Hazardous Materials and Non-Hazardous Materials Truck Shipment Accidents/Incidents.* http://www.phmsa.dot.gov/pv_obj_cache/pv_obj_id_2F1F1A79D0F916679838CDE369E5600B20733400/filename/hazmatriskfinalreport.pdf.

DOT (Department of Transportation). 2004. *Crashes Involving Trucks Carrying Hazardous Materials.* FMCSA-RI-04-024. Federal Motor Carrier Safety Administration. May. http://www.fmcsa.dot.gov/facts-research/facts-figures/analysis-statistics/fmcsa-ri-04-024.htm.

DOT. 2013. *Large Truck and Bus Crash Facts–2011.* FMCSA-RRA-13-049. Federal Motor Carrier Safety Administration. October. http://www.fmcsa.dot.gov/safety/research-and-analysis/large-truck-and-bus-crash-facts-2011.

DOT. 2014a. PHMSA Incident Statistics. http://www.phmsa.dot.gov/hazmat/library/data-stats/incidents. Accessed August 7, 2014.

DOT. 2014b. Company Snapshot: Bed Rock Inc. http://safer.fmcsa.dot.gov. Accessed August 1, 2014.

Guelta, M.A., and L. Fazekas-Carey. 2003. *Biodegradation of Hydrolyzed Mustard from an Assembled Chemical Weapons Assessment (ACWA) Projectile Washout Study.* ECBC-TR-291. Aberdeen Proving Ground, Md.: Edgewood Chemical Biological Center.

JPEOCBD (Joint Program Executive Office for Chemical and Biological Defense). 2014. *Offsite Hydrolysate Transport and Disposal Case Studies.* Provided by PCAPP on July 30, 2014.

Noblis. 2008. *Offsite Disposal of ACWA Hydrolysates.* NTR 2008-61129. Falls Church, Va.

NRC (National Research Council). 2008. *Review of Secondary Waste Disposal Planning for the Blue Grass and Pueblo Chemical Agent Destruction Plants.* Washington, D.C.: The National Academies Press.

PMACWA (Program Manager, Assembled Chemical Weapons Assessment). 2006. *Questions & Answers—Hydrolysate.* Available at the Pueblo Chemical Stockpile Outreach Office. Pueblo, Colo.

PMACWA. 2012. *PCAPP Hazardous Materials and Waste In and Out of the Facility.* Provided by PCAPP on August 18, 2014.

RCMD (Recovered Chemical Materiel Directorate). 2014. *Bomblet Destruction Campaign at Rocky Mountain Arsenal.* Provided by PCAPP on July 30, 2014.

TRB (Transportation Research Board). 2013. *Hazardous Materials Transportation Risk Assessment: State of the Practice,* Visual Risk Technologies, Inc., Hazardous Materials Cooperative Research Program Report 12. Washington, D.C.

6

Hydrolysate Treatment Performance Goals

All stakeholders in the Pueblo Chemical Agent Destruction Pilot Plant (PCAPP) project, including Pueblo community members, the Colorado Citizens' Advisory Commission (CAC), PCAPP government and contractor employees, Assembled Chemical Weapons Alternatives (ACWA) management, the Colorado Department of Public Health and Environment (CDPHE), and the U.S. Departments of Defense and State, share a common goal: to safely and efficiently destroy the munitions stored at the Pueblo Chemical Depot (PCD). Chemical weapons have been stored at the PCD for over 60 years. Based on discussions with PCAPP officials and CAC members, community perceptions of the risks posed by the PCD appear to be relatively stable.[1] As weapons are destroyed at PCAPP and the project approaches completion, this risk profile will diminish over time toward zero. Delays in the destruction process would halt risk reduction and could protract the community's risk profile. It can even be argued that delays would increase the risk posed by the stockpile, owing, for example, to aging munitions or loss of trained personnel.

The long-standing presence of the stockpile may lead community members to perceive the existing risk as a baseline for decision making. If this is the case, new risks that may arise from PCAPP operations—for example, shipments of hazardous materials on- and offsite, odors, and emissions—could be perceived as increasing the risk to the community. Risk-based decision making should balance the new risks against the larger, existing risk posed by the longtime existence of the stockpile. This section discusses criteria for success in treating hydrolysate with this broader risk picture in mind.

[1] Conference call on September 5, 2014, with Judith Bradbury and Hank Jenkins-Smith, members, Hydrolysate Committee; Irene Kornelly, chair, Colorado CAC; Jeannine Natterman, CDPHE; John Norton, member, Colorado CAC; Sandy Romero, PCAPP communications manager; Nancy Schulte, study director, Hydrolysate Committee; and Thomas Schultz, public affairs specialist, PCAPP.

DECISION FRAMEWORK FOR DETERMINING SUCCESSFUL PLANT OPERATION

PCAPP project stages include (1) design, (2) construction, (3) systemization, (4) pilot testing, (5) operation, and (6) closure.[2] The overall systemization plan prepares PCAPP for success in three areas: paper, plant, and people. The committee addresses primarily the risk reduction and mitigation portion of the "plant" subset of the overall systemization plan (PCAPP, 2014).

PCAPP is piloting a number of treatment steps that will be applied to the unique waste streams stemming from the production of hydrolysate. As a result, there are uncertainties in how well the end-to-end process will perform. In particular, the biotreatment and brine reduction systems have never been utilized for a substrate like the hydrolysate feed. Based on previous biodegradation testing, system modeling, and analysis, there is a high probability that the required level of thiodiglycol (TDG) biodegradation can be achieved. However, the efficiency and effectiveness of the PCAPP system will remain unknown until the system is placed in operation with actual hydrolysate feed.

To minimize risks associated with new applications of these treatment processes, risk reduction and mitigation activities are identified and addressed through a series of planned preoperational testing activities (PCAPP, 2014). Data from these preoperational tests will help to determine the likelihood that the system will meet performance criteria and will identify and test process alternatives in the case of process underperformance or failure.

The primary criteria for successful treatment of hydrolysate are (1) meeting Resource Conservation and Recovery Act (RCRA) permit requirements and (2) meeting ACWA requirements for treatment of the hydrolysate. This includes the production of process water from the brine reduction

[2] See the website for the Program Executive Office, Assembled Chemical Weapons Alternatives, at http://www.peoacwa.army.mil, accessed January 29, 2015.

system (BRS) that is of good enough quality that it can be recycled to the plant and production of a suitable filter cake that meets regulatory requirements for offsite disposal. In addition, it will be desirable to meet ACWA's schedule for destruction of the munitions at PCAPP. In this manner, overall risk to the community in and around PCD and PCAPP will be gradually reduced. In other words, issues with operation of the biotreatment area (BTA) may be overcome while still complying with RCRA and ACWA requirements; however, they should not delay the overall schedule for destruction of the munitions.

If the plant is unable to satisfy these criteria and if the time it takes to destroy the actual munitions by PCAPP increases as a result of degraded performance of the BTA, the risk reduction goals associated with destruction of the stockpile will not be achieved in a timely manner. If the schedule for overall destruction of the munitions is delayed due to irreparable issues with the BTA, the best option for continued risk reduction to the community will be to ship the hydrolysate offsite for treatment and disposal.

Finding 6-1. The primary criteria for successful treatment of hydrolysate are (1) meeting RCRA permit requirements and (2) meeting ACWA requirements for treatment of the hydrolysate. This includes production of process water from the BTA that is of high enough quality that it can be recycled to the plant and production of a filter cake that meets regulatory requirements for offsite disposal. In addition, it will be desirable to meet ACWA's schedule for destruction of the munitions at PCAPP.

The overarching question posed to the committee is this: What are PCAPP's options to accomplish its mission of munitions demilitarization if the biotreatment and/or brine reduction systems underperform? The committee considered underperformance of the overall system as failure to meet permit requirements and failure to adequately remove specific components from the hydrolysate feed, such that the liquid effluent no longer meets requirements for reuse within the plant.

While the critical decision from the perspective of most stakeholders is if/when offsite shipment of hydrolysate becomes a necessity, a broader evaluation of PCAPP's operational options must first be performed. Performance goals for each unit process will have to be evaluated, and a decision will need to be made regarding alternative options. If alternative options still fail to treat the hydrolysate to an adequate level, the decision to consider offsite shipment of the hydrolysate would become more likely.

The decision framework illustrated in Figure 6-1 captures, at a high level, the considerations and decisions that drive major changes to operations. More specifically, this decision framework allows decision makers to evaluate all available options before resorting to offsite shipment of hydrolysate.

The PCAPP BTA treatment process involves three main unit processes—immobilized cell bioreactors (ICBs), the water recovery system (WRS), and the BRS—that must meet defined process performance goals in order to ensure overall system performance. The individual performance goals for these processes are evaluated in Chapter 7 using the decision framework illustrated in Figure 6-1. This simple decision framework for evaluating system performance based on performance criteria will aid in reducing risk of underperformance by helping to define reasonable alternative operational decisions. The decision framework shown here will also help build consensus among stakeholders and decision makers on important decisions about major changes to PCAPP operations; this framework is applied to individual unit processes in Chapter 7.

As described at the outset of this chapter, there are multiple stakeholders invested in the success of the PCAPP project. However, it is important to note that not all stakeholders are decision makers when it comes to making changes in operations. The PCAPP operations staff is responsible for the operations of the plant. As a result, they are the initial set of decision makers on routine, day-to-day adjustments to operations. For example, if problems are encountered in maintaining the pH or nutrient levels in the ICBs, potentially causing the system to underperform, PCAPP staff would take corrective actions to return these levels to an acceptable operational range. If an issue like this persists, the decision to pursue further alternatives might be elevated to PCAPP management or to the ACWA level and might include decisions on changes to infrastructure, especially those that might incur costs or delays—for example, installing extra holding tanks to store hydrolysate feed—or necessitate permit modifications. Additional stakeholders may therefore need to be consulted in these decisions depending on their nature or the magnitude of their impact.

PERFORMANCE CRITERIA FOR HYDROLYSATE

The decision framework is supported by a set of performance criteria (indicated by red arrows in Figure 6-1). These criteria, which include effectiveness, efficiency, cost, regulatory/permitting requirements, and stakeholder agreement, are used at two points in the framework. This enables decision makers to compare the performance of the existing system operations with the objectives or perhaps with alternative system operations. A set of performance criteria would include those outlined in Box 6-1. A goal or threshold for each criterion would need to be established by decision makers to define "successful" operations. The use of multiple criteria for performance evaluation allows decision makers to document the extent of underperformance. In other words, while a total failure of the system (although very unlikely) may be easily identified, a partial failure or a temporary failure and the resulting impacts may be more difficult to define, to predict, and to address. Multiple per-

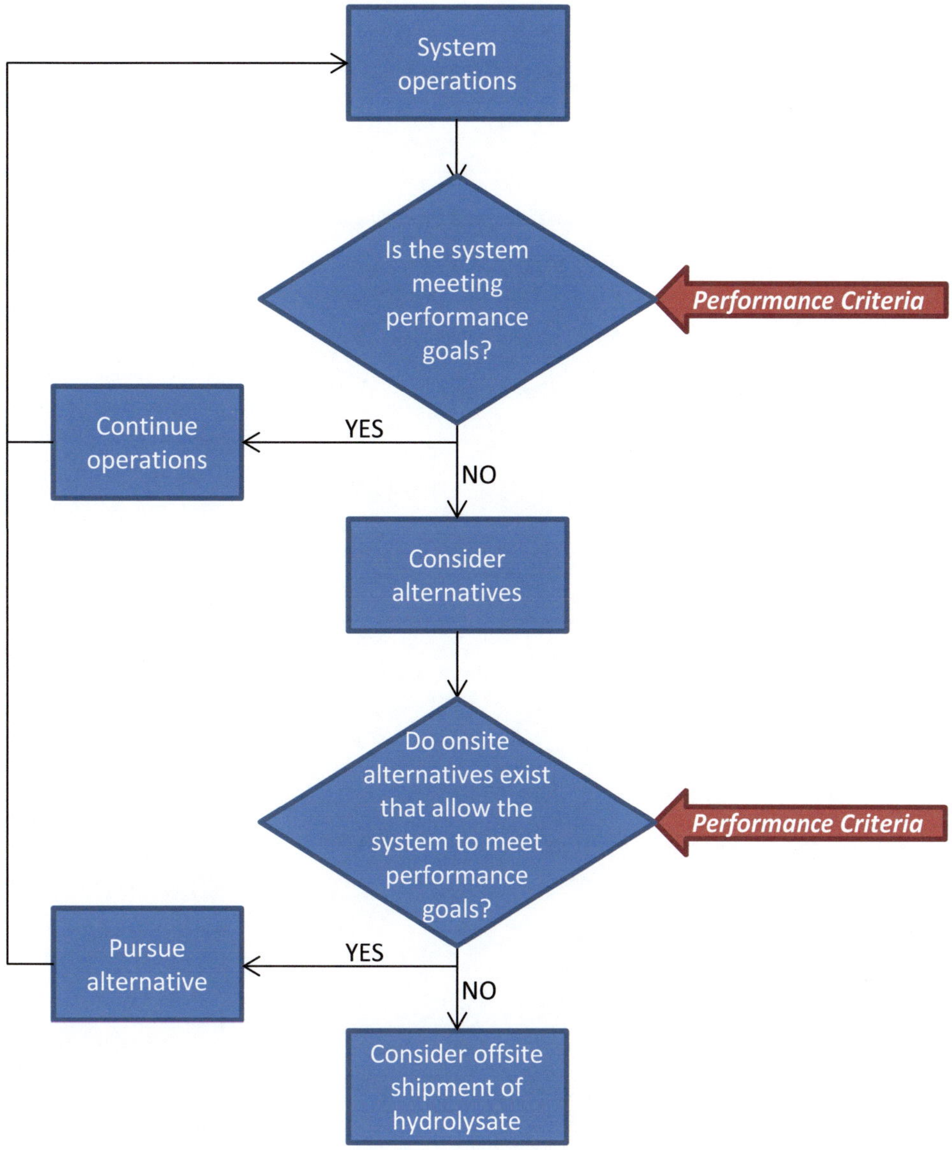

FIGURE 6-1 Decision framework.

formance criteria add clarity to these more likely cases of underperformance.

To apply the decision framework, technical factors that could cause the system to underperform are identified. For example, if there is insufficient oxygen to meet the demand of the biomass, then the extent of TDG biodegradation is decreased. The impact of this technical factor is then evaluated to determine if operational changes should be considered. In the same example, limited oxygen supply might impact multiple criteria, including targets for biodegradation, recycle water quality, and permitting requirements. If the system is underperforming based on an assessment of these criteria, alternative onsite options (i.e., contingency strategies) are then considered. In this case, modifying the influent piping to distribute the feed across multiple chambers may improve the distribution of oxygen to meet biomass needs. As before, the impacts of implementing this alternative can be evaluated using the same performance criteria. This alternative, although it would improve the effectiveness criteria, would incur infrastructure costs and might impact the project schedule. A similar process can be used to evaluate other contingency strategies in order to determine how to proceed. At each decision point, the relevant benefits and liabilities of each option would be carefully considered.

Once multiple technical factors and contingency options have been evaluated using the performance criteria, the overall system risk can be assessed holistically. A graded evaluation of system risk would allow stakeholders to qualitatively rate the potential for overall program success at any point in the project. This type of graded evaluation will

BOX 6-1
Performance Criteria for Hydrolysate

Effectiveness Criteria

- Level of TDG biodegradation
  —Design treatment goal of ≥95 percent at design flow rate, or ≥98 percent at 50-90 percent flow, TDG destruction in the ICBs.
  —Minimum treatment goal of <0.1 wt% TDG when the ICB unit is being air scoured or in recovery. This would result in a treatment of >86 percent.[a]
  —PCAPP RCRA RD&D permit performance goal of >95 percent and a target of at least 99 percent removal of the TDG and at least an average of 90 percent removal of TOC.[b]
- Filter cake must meet offsite TSDF acceptance criteria.
- Recycle water must meet treatment requirements so as to allow reuse within the plant.
- Odor and emissions:
  —Odor levels should not exceed a dilution-to-threshold ratio (D/T) of 7/1.[c]
  —Emissions should comply with the release requirements specified in the RCRA permit.

Efficiency/Schedule Criteria

- The BTA process rate should meet or exceed the agent neutralization process rate. Throughput buffer capacity = 30 days for ICB feedstock, plus storage capacity of WRS.
- Project schedule and projected completion date:
  —Munitions destruction projected to be completed according to the ACWA schedule.
  —Extended delays (e.g., beyond 30 days) may result in loss of trained workforce.
  —The length of "acceptable" delay may vary across stakeholders.

Cost Criteria

- All operating costs (including unanticipated maintenance, infrastructure, and other costs) are within available funding.

Regulatory/Permitting Requirements

- CDPHE RCRA permit.
- Pueblo County certificate of designation.
- National Environmental Policy Act.
- Organisation for the Prohibition of Chemical Weapons.

[a] Paul J. Usinowicz, Bechtel technical advisor, "Biotreatment Area Cradle to Grave," presentation to the committee, July 30, 2014.

[b] PCAPP RCRA RD&D Stage III, Class 3, Permit Modification Request, Revision 0, November 2006, Appendix D, Waste Analysis Plan, C-2a(3)(d), p. C-11.

[c] The dilution-to-threshold (D/T) ratio is a measure of the number of dilutions needed to make the odorous ambient air nondetectable. It is calculated as (Volume of carbon-filtered air)/(Volume of malodorous air). See RCRA permit modification #31, BTA and BRS Odor Control and Colorado Regulation No. 2, 5 CCR 1001-4. Note that the limit shown in the *White Paper Bio-Treatment Area Risk Reduction and Mitigation* (PCAPP, 2014), p. vii, is an odor-to-threshold ratio of 1.0 or greater.

facilitate communication between stakeholders and allow stakeholders to track and document PCAPP progress in a transparent and consistent way throughout the course of the project. Table 6-1 exemplifies a graded scale for success that could be used for the PCAPP project.

Finding 6-2. The PCAPP systemization plan (including the risk reduction and mitigation plan) lacks a decision-making process to guide major changes in operations.

Recommendation 6-1. PCAPP should implement a clear decision-making process to guide major changes in operations.

In Chapter 7, potential technical factors in the ICBs, WRS, and BRS that may lead to insufficient treatment of hydrolysate are identified. For each potential factor, the impact on system performance is described and alternative operational strategies are identified. Finally, the technical factors, along with their corresponding alternatives, are assessed using the graded scale described in Table 6-1. The committee judges

TABLE 6-1 Graded Success Scale for Use in Evaluating Overall Operation and Individual Treatment Processes (ICBs, WRS, and BRS)

Grade	Definition
0	Success is practically certain (very low possibility of project failure): Operations are proceeding as expected. No PCAPP actions needed.
1	High likelihood of success (low possibility of project failure): Actions should be taken by PCAPP to prepare ahead of time for implementation of contingencies in the event of failures. For example, PCAPP might begin to prepare permit modifications and planning documents, including building plans for piping and shipping.
2	Success is uncertain (moderate possibility of project failure): Actions should be taken to prepare for implementation of contingency operations. For example, PCAPP might begin processing environmental documentation and finalizing contingency plans, purchasing needed materials, and implementing changes to the infrastructure.
3	Success is unlikely with current operations (high possibility of failure of the project): Actions are taken to accelerate the implementation of contingency operations. For example, construction of needed facilities is completed as quickly as possible, and environmental approvals are expedited if they have not already been obtained.

NOTE: WRS, water recovery system; BRS, brine reduction system.

that PCAPP is currently operating within Grade 1. Although there is a high likelihood of success, actions should be taken by PCAPP to prepare ahead of time for implementation of contingencies in the event of failures.

Finding 6-3. The committee judges that PCAPP is currently operating within Grade 1. There is a high likelihood that PCAPP will be successful in meeting performance criteria. Conversely, there is a low possibility of failure.

Recommendation 6-2. Actions should be taken by PCAPP to prepare ahead of time for implementation of contingencies in the event of failures.

A number of the technical factors identified in Chapter 7 are being evaluated during the systemization, risk reduction, and surrogate testing processes. In this way, alternative operational strategies may be implemented before agent operations begin.

PREOPERATIONAL TESTING DATA THAT FACTOR INTO THE DECISION PROCESS

Systemization will provide PCAPP the opportunity to identify potential failure points, test alternative operational and treatment strategies, and provide valuable plant operating experience. Systemization planning should reduce negative impacts on the overall schedule due to unanticipated operational problems and reduce overall operating costs. In addition, the "Risk Reduction and Mitigation Plan" specifies additional testing beyond the previously anticipated systemization testing (Guzzetti, 2014). The systemization plan consists of approximately 6 months of operations, with the first 2 to 3 months dedicated to start-up/acclimation needs and development of steady-state operational conditions. The remaining 3 to 4 months will allow the system to operate under steady-state conditions to assure system stability and allow for data to be collected and used to assess performance and evaluate alternatives (PCAPP, 2014). Careful data collection, performance evaluation, and corrective mitigation activities taken during the start-up and steady-state phase of systemization should decrease the likelihood of delays or system failure during actual system operation.

PCAPP has outlined a risk reduction and mitigation plan that will evaluate many of the technical factors identified in the next chapter that could lead to underperformance of the BTA. Table 6-2 briefly summarizes the data that will be collected during the risk reduction and mitigation effort that will be performed concurrently with systemization. These data are being used to inform operational changes and adjustments that may need to be made before agent treatment begins in September 2015. Once treatment of actual munitions begins, operational data will be collected to monitor performance of the overall system. Operational data can be used to assess performance against the established criteria for success and will therefore be critical to the decision-making process. In the event that the BTA still fails to meet performance goals despite the use of alternative operational strategies, the downstream impacts of offsite shipment will need to be considered. These issues are addressed in Chapter 7.

Finding 6-4. In its recent white paper on risk reduction and mitigation, PCAPP has done a thorough job of identifying potential failure risks and providing targeted strategies to mitigate these risks in the BTA (PCAPP, 2014). Employing the decision-making framework outlined previously, the overall systemization plan, and the BTA risk reduction and mitigation plan provides targeted strategies for PCAPP to mitigate any operational problems that become apparent during surrogate testing and systemization.

TABLE 6-2 Description of How Data Generated During Systemization Could Be Used to Reduce Risk of Ineffective Operation or System Failure, and Possible Alternatives to Reduce These Risks

Systemization Phase	Type of Data Generated	Implication/Use	Contingency Option
Enhanced water testing in the BRS	• pH • Temperature • Water throughput rate	Assess the ability of the vendor's equipment to operate as planned.	Replace equipment that does not meet specifications.
Respirometry studies	• TDG biodegradation rate for various biomass seed • Nutrient consumption • Acid production TDG toxicity	Determine expected operational conditions, preferred biomass source to seed the ICBs, and TDG concentration for initial start-up phase.	Utilize alternative biomass seed source and modify TDG concentration for start-up phase.
Surrogate testing (ICBs)	• Percent biodegradation of TDG under site-specific conditions • pH and temperature • DO level in ICBs • Monitor for inhibitory TDG concentrations (start-up) • Nutrient consumption and accumulation • Biomass accumulation	Control of operational parameters within the ICBs is critical to degradation of TDG.	• Modify operational conditions to meet pH thresholds. • Modify pH control systems. • Replace or update monitoring equipment. • Modify TDG loading for start-up and/or distribute feed to more than one chamber. • Determine need for cooling system within ICBs. • Modify procedures for nutrient control. • Evaluate alternative options for hydrolysate disposition.
Surrogate testing (WRS)	• Aeration • Odors • Biosolids concentration	Ensure that aeration and mixing are adequate.	• Modify holding tanks to ensure mixing and aeration. • Modify solids removal equipment.
Surrogate testing (BRS)	• Foaming • Fouling • Water content in filter cake • Water quality of effluent	Effluent from BRS must meet permitting requirement for both liquid (recycle) and solid (shipment offsite) components.	• Add antifoaming agents. • Add solids removal process prior to BRS. • Evaluate requirement for disposal of filter cake. • Evaluate possibility of recycling some portion of water. • Evaluate possibility of offsite shipment.

SOURCE: Guzzetti, 2014.

REFERENCES

Guzzetti, D., start-up field supervisor, PCAPP, and D. deLesdernier, Battelle, support. 2014. "Biotreatment Area Risk Reduction and Mitigation," presentation to the committee on July 29, 2014.

PCAPP. 2014. *White Paper Bio-Treatment Area Risk Reduction and Mitigation*. 24852-30H-BTA-V0001. Pueblo, Colo.: Pueblo Chemical Agent Destruction Pilot Plant.

7

Failure Risks, Systemization, and Contingency Options

This chapter discusses the possible risks of failure within the biotreatment area (BTA) at the Pueblo Chemical Agent Destruction Pilot Plant (PCAPP). Figure 7-1 shows the three main subsystems of the BTA—immobilized cell bioreactors (ICBs), the water recovery system (WRS), and the brine recovery system (BRS)—and the various inputs and outputs of the overall system. PCAPP identified several sources of potential failure, along with contingency options, in its recent white paper on risk reduction and mitigation in the BTA (PCAPP, 2014). It plans to evaluate the majority of these risks during systemization. This chapter expands on the analysis in the white paper by considering the risk of failure once the system begins treating actual munitions and focuses on decisions leading to changes in plant operations. The decision framework, the performance criteria, and the graded scale for success that were introduced in Chapter 6 are leveraged in the discussion.

This chapter is organized in the following manner. To simplify the evaluation, the ICBs and the BRS are considered separately and the WRS is considered to be a part of the BRS. For each component, the following topics are considered:

1. Technical factors that may lead to insufficient treatment.
2. Systemization and likelihood of insufficient treatment.
3. Impacts if system underperforms or does not perform.
4. Contingency options for onsite operations.
5. Summary table with graded performance.

Next, the unlikely scenario of multicomponent failure is discussed. If all contingency options are deemed ineffective, or if multicomponent failure or catastrophic failure occurs, decision makers may be led to shift operations toward offsite shipment of hydrolysate. Specifically, this section considers actions that must be taken to prepare for and implement offsite shipment of hydrolysate. Finally, the chapter concludes with a comparison of the Colorado Citizens' Advisory Commission (CAC) concerns (reprinted in Appendix A) with the risks and options discussed herein.

FAILURE RISKS, SYSTEMIZATION, AND CONTINGENCY OPTIONS IN THE IMMOBILIZED CELL BIOREACTOR

As discussed in Chapter 2, there are several advantages to using the ICBs to implement biotreatment for the unique and complex hydrolysate feed. The aerobic biodegradation of thiodiglycol (TDG) will yield primarily mineralized products of water, carbon dioxide, and sulfuric acid. The planned steady-state performance within the ICBs is intended to destroy at least 95 percent of the TDG in the hydrolysate feed. The following section discusses the failure risks and contingency options in the ICBs.

Technical Factors That May Lead to Insufficient Treatment

Several technical factors may lead to insufficient treatment of TDG in the ICBs, especially under off-normal conditions. For example, although TDG is readily biodegradable under aerobic conditions, it is toxic to the degrading biomass at elevated TDG concentrations. That is the reason for diluting the hydrolysate from the agent processing building by a factor of 8 (one part hydrolysate to seven parts process water) with recovered process water from the BRS. Thus, if there is incomplete biodegradation of the TDG in the ICBs and the concentration of TDG increases, toxic inhibition can slow down the biodegradation rate, which further promotes the accumulation of TDG in the bioreactor.

In addition to toxicity concerns from the TDG itself, incomplete biodegradation of TDG can result if the pH is not in the optimum range of 7 to 8. Both higher and lower pH values will slow biodegradation. The hydrolysate generated from the caustic neutralization of mustard agent is at pH $\geq$10, so there is a provision to add sulfuric acid in the ICB's feed tank to lower the pH of the feed water to the ICBs. The bio-oxidation of TDG in the ICBs is expected to yield sul-

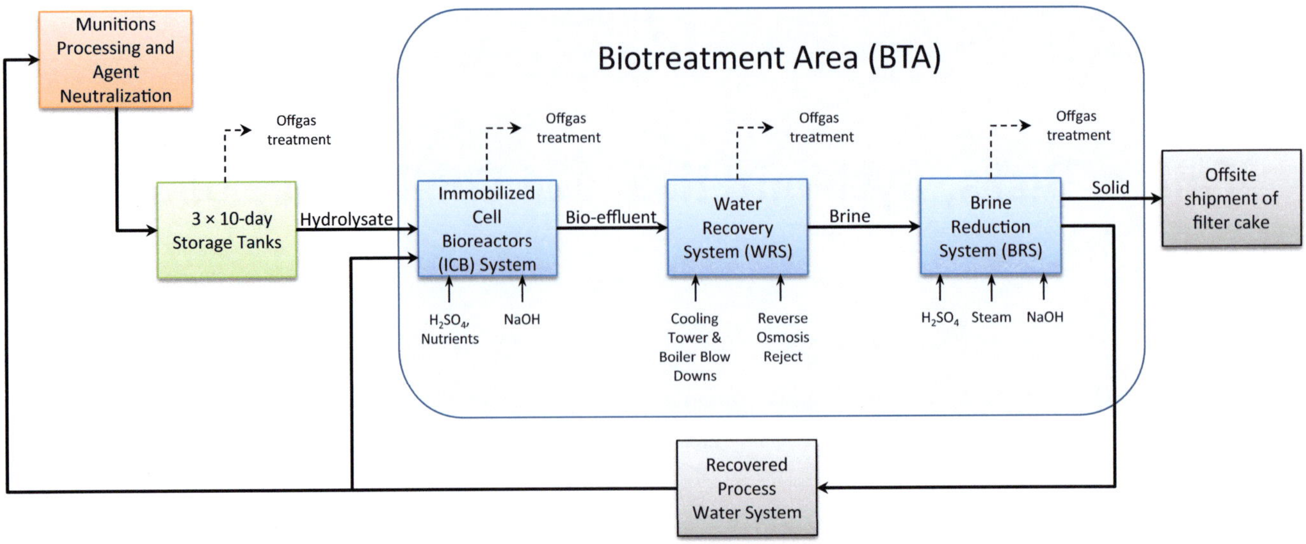

FIGURE 7-1 PCAPP BTA process diagram. SOURCE: Adapted from Don Guzzetti, start-up field supervisor, PCAPP, and Dave deLesdernier, support, Battelle, "Biotreatment Area Risk Reduction and Mitigation," presentation to the committee on July 29, 2014.

furic acid, which will tend to lower the pH in the treatment units. Consequently, provision has been made for sodium hydroxide (caustic) to be added to the ICBs to help raise the pH during biotreatment. Because control to near-neutral values is difficult to achieve by adding a strong acid or strong base, fluctuations in the pH could be an issue for effective operation of the ICBs. The addition of sodium carbonate (soda ash), which is a weaker base, could be considered as an alternative to sodium hydroxide to help buffer the pH to between 7 and 8.

Operating temperature influences biodegradation performance. A previous study that examined the removal of TDG found that biotreatment was optimal between 8°C and 35°C (46°F and 95°F) and that the performance significantly declined when biotreatment was conducted at 35°C to 40°C (95°F to 104°F) (SBR Technologies, Inc., 1996). The ICBs are equipped with steam injection ports for heating during cold weather operation, but there are no provisions for cooling the ICBs. If the summer heat and solar radiation cause the temperatures in the ICBs to exceed the optimum range, then the TDG biodegradation will likely slow down and contribute to incomplete treatment.

In light of the potential toxicity and inhibition issues, acclimation of the biomass during start-up will be critical, as start-up represents a non-steady-state condition. If transition from the planned operations with the simulated hydrolysate to operations with the actual 1:8 diluted hydrolysate feed takes longer than scheduled due to slow biomass acclimation, this can delay the ability to treat the hydrolysate stream being generated from the munitions processing and agent neutralization step. Hydrolysate storage in the three 10-day tanks does provide some time cushion if biomass acclimation takes longer than scheduled or time is needed for recovery of the biotreatment process. If a delay is longer than 30 days, then hydrolysate can no longer be generated, and the munitions processing must be temporarily halted unless contingency plans are in place that can be implemented rapidly. The composition of the hydrolysate is anticipated to be relatively uniform, and the hydrolysate in the feed tanks will be diluted and acid-neutralized to produce a uniform feed concentration. These two factors are expected to promote steady-state conditions for the ICBs if the pH and temperature can be adequately controlled.

Two other factors can slow the TDG biodegradation rate: nutrient and oxygen limitations. The hydrolysate is devoid of adequate nutrients to support biomass growth. Both nitrogen and phosphorus must be added to the hydrolysate feed tanks as supplemental nutrients. Nitrogen will be added as urea and phosphorus as diammonium phosphate (DAP). The iron in the hydrolysate might cause $FePO_4$ to precipitate, which will reduce the availability of phosphorus for the biomass, further inhibiting TDG biodegradation. Air will be supplied to the ICBs via coarse diffusers to provide oxygen. The air supply is evenly distributed throughout the three chambers, with all the TDG loading in the first chamber. The greatest demand for oxygen will therefore occur in the first chamber, and, if the demand exceeds the supply, there will not be enough oxygen for the biomass. Too little oxygen will lead to anaerobic reactions, which typically proceed at slower rates (Metcalf and Eddy/AECOM, 2014) or to reactions in which sulfates are reduced to sulfides. Sulfide concentrations ≥100 mg/L can be toxic to the bacteria (NRC 2013, p. 15). Both processes lead to less effective destruction of TDG.

Excessive biomass growth and accumulation of other solids in the ICBs could hinder the TDG treatment efficiency by way of at least two mechanisms. First, the accumulation of solids could clog the support medium and hinder the flow of wastewater through the immobilized biomass, leading to channeling and short-circuiting of the flow, reducing the contact time with the biomass, and reducing the extent of TDG biodegradation. Second, if large quantities of biomass grow and accumulate on the support medium, periodic sloughing events could lead to extensive short-term biomass loss, thereby decreasing the amount of TDG that can be biodegraded in the period immediately following the sloughing event and until new biomass accumulates on the support medium. Such periodic sloughing events will cause spikes in the solids loading to the BRS and may impact operations downstream by potentially fouling heat transfer surfaces, producing acid upon hydrolysis of the biomass at the high temperatures of the evaporator and increasing the organic content of the solids produced in the crystallizer.

One additional concern is the release of malodorous compounds during aeration. The system is designed with granular activated carbon (GAC) adsorbers to capture volatile chemicals, including vinyl chloride, that are present in the hydrolysate or that are generated during the course of biodegradation. Some of these compounds may also be malodorous. If these odorous compounds are not adequately removed, they may pose a problem with one of the performance goals—for example, exceeding the dilution to a threshold (D/T) ratio of 7 at or beyond the fence line.[1]

Finding 7-1. The hydrolysate generated by agent neutralization is a complex mixture. While ICB reactors have been used successfully to treat complex hazardous organic wastes, they have not been used to treat this specific hydrolysate.

Finding 7-2. Technical factors that may lead to insufficient hydrolysate treatment include hydrolysate toxicity to microbial biomass, the need for careful pH and temperature control, nutrient and oxygen limitations, biomass buildup and sloughing, start-up and acclimation issues, and release of malodorous compounds.

Recommendation 7-1. PCAPP should develop contingency plans to mitigate risk in the event that one or more of the above factors inhibit efficient ICB operations. Such plans should be in place prior to system start-up so that agent neutralization operations are not unduly delayed.

[1] RCRA Permit modification #31, BTA and BRS Odor Control and Colorado Regulation No. 2, 5 CCR 1001-4. Note that the limit shown in PCAPP (2014), p. vii, is an "odor-to-threshold" ratio of ≥1.0.

Systemization and Likelihood of Insufficient Treatment

Although the hydrolysate is a complex feed that has not been treated before with ICBs under the planned full-scale operating conditions, there is a high probability that successful biodegradation of the TDG can be achieved at PCAPP. First, bench-scale biodegradation tests and pilot-scale bioreactor studies have demonstrated that effective biological destruction of TDG can be achieved (NRC, 2013). Second, because the hydrolysate feed to the ICBs is expected to be relatively uniform in composition, typical problems associated with fluctuating organic loads and flows should be avoidable and the ICBs can be anticipated to exhibit steady-state behavior. Third, the immobilized biomass provides greater resistance to fluctuations in organic loading (should they occur), exposure to toxic compounds, and faster recovery from excursions in pH and temperature than suspended growth bioreactors. Fourth, pH monitoring and control capabilities have been provided so that the ICBs can be operated under their optimal pH conditions. Fifth, under normal operations, there is excess hydraulic capacity within the ICBs relative to the anticipated production flow rate of the hydrolysate, so the ICBs are not expected to be the rate-limiting step in the overall mustard agent destruction process. Separation of the agent neutralization process from the downstream hydrolysate biotreatment allows each of these two operations to be operated with considerable independence. The 30-day storage capacity for the hydrolysate provides a cushion to minimize the impact of potential delays and downtimes on the operations schedule.

Finally, the risk reduction and mitigation efforts during systemization will test a number of key process variables, identify potential failure points, and determine optimal ways to operate the downstream processes at PCAPP (PCAPP, 2014). For example, one phase of the systemization studies will be conducted with an ICB unit filled only with water to measure the temperature of the ICBs during the hot days of summer and maximum solar heating, evaluate the ability to control pH through caustic or acid addition, and test the recycle piping and pumps for the process dilution water. A second study will involve respirometry to screen for an appropriate biological seed from local municipal wastewater treatment plants, assess acclimation requirements for the selected biological seeds, determine the maximum hydrolysate loading, and compare the biodegradation response to the surrogate hydrolysate feed and to samples of actual hydrolysate. Respirometry measures the amount of oxygen consumed in a batch culture over time as an indicator of the rate and extent of the biodegradation and can also be used as an indicator of microbial inhibition and toxicity.

In the last phase of systemization, a subset of the ICBs will be operated for 4 months with a surrogate hydrolysate feed containing actual TDG. This program appears to be well thought out and designed. PCAPP has purchased TDG and will prepare a surrogate hydrolysate in the 30-day storage tanks that contains 5.6 wt% TDG, 5.4 wt% NaCl, 0.2 wt% $FeCl_3$, and caustic to raise the pH to 10. The ICBs will be seeded using the biomass with the best TDG biodegradation properties as determined from the respirometer studies. A 2-month start-up period is planned during which the TDG loading will be gradually ramped up to the design loading, followed by 2 months of steady-state operation with the surrogate hydrolysate as the feed. This testing with the surrogate hydrolysate provides an opportunity to assess the start-up process, demonstrate the effectiveness of TDG removal, confirm aeration and nutrient requirements, demonstrate adequate pH control, assess the extent of solids accumulation and their impacts, confirm the heat-transfer model for simulation of operating temperatures, and gain operational experience with the biotreatment and biomass activity. It will also help with downstream evaluation of the WRS and BRS. The timing of this testing with surrogate hydrolysate will dovetail with the start-up of the actual agent neutralization operations, thus allowing for a blended transition from surrogate hydrolysate to actual agent hydrolysate. This will allow full-scale operation of the ICBs to commence with an acclimated biomass and will facilitate rapid start-up and transition to steady-state operation of the ICBs treating hydrolysate.

Finding 7-3. There is a high likelihood that successful biodegradation of the TDG can be achieved, but there is still a small possibility that the overall performance of the BTA will not meet expectations.

Finding 7-4. The proposed risk reduction and mitigation plan will test a number of key process variables, identify potential failure points, and determine better ways to operate the downstream processes at PCAPP.

Finding 7-5. Sequencing of the risk reduction and mitigation effort with surrogate hydrolysate and full-scale hydrolysate generation and biotreatment operations should allow for timely start-up and steady-state operation of the ICBs.

Impacts If ICB Systems Underperform or Do Not Perform

If the ICBs provide substantial biodegradation of the TDG but are not able to achieve the design treatment goal of at least 95 percent TDG destruction, the downstream BRS may still be able to separate the remaining TDG from the liquid effluent and produce water of adequate quality for recycling. The impact would be increased amounts of TDG in the solid cake that is shipped offsite for disposal and, possibly, higher TDG loadings to the GAC adsorbers that treat the product water from the BRS. The additional organic loading on the BRS will probably mean more frequent replacement of the GAC or require more GAC adsorption capacity. Poor TDG removal in the ICBs could overload the capacity of the BRS and cause it to fail. If this happens, the water produced may be of inadequate quality for recycling and may not meet the anticipated permit requirements. Note also that if the ICBs destroy a substantial amount of the TDG but are not able to achieve the design treatment goal of at least 95 percent TDG destruction, mitigative measures could also entail the need for permit modifications and discussions with the Organisation for the Prohibition of Chemical Weapons (OPCW) regarding TDG destruction efficiency, both of which could lead to a significant delay in agent processing.

If the combination of the ICBs and BRS is not able to yield process water that can be recycled, then a means to dispose of the BRS effluent must be found. Because the hydrolysate is diluted 1:7 to reduce TDG toxicity of the feed to the ICBs, a much larger volume of water will need to be disposed of. If the ICB and BRS combination is not able to produce water that is adequate to be recycled as process water, then a decision may be required on whether it would be more prudent to ship the undiluted hydrolysate offsite for treatment rather than to attempt to dispose of the much larger volume of BRS effluent. This decision should be based on an evaluation of the criteria for success and the agreed-upon decision framework described in Chapter 6.

Contingency Options

A number of alternative operational strategies (i.e., contingency options) may be implemented if insufficient biotreatment of TDG is realized with the ICBs or if operational problems are encountered. Such alternative operational strategies as those described below aim to maintain onsite processing of hydrolysate in the BTA. Some performance limitations may become apparent during the risk reduction and minimization studies, so that timely action can be taken to avoid delays in agent neutralization operations.

- The TDG loading can be reduced to overcome toxic inhibition and decrease the demand for oxygen. The flow rate to the ICBs can be reduced to increase the hydraulic residence time. This modification will result in less hydrolysate treated in a given time period compared to the designed operating capacity, but the performance may still be adequate.
- If $FePO_4$ precipitation is responsible for the insufficient treatment, then moving the feed point for the DAP from the ICB feed tank to the recycle line for each ICB unit can be advantageous. Adding DAP to the feed tank will likely result in more $FePO_4$ precipitation because of the higher residence time in the feed tank, between 10 and 30 days, and increased iron availability. Adding the DAP to the recycle line will lessen the time of contact of the DAP with the iron originally in the hydrolysate and diminish the amount of $FePO_4$ that precipitates. This will also improve the availability of phosphorus for biomass growth.

- If pH cannot be adequately controlled through the addition of strong acid (sulfuric acid) and strong base (sodium hydroxide), consideration should be given to using sodium carbonate as a weaker base to buffer the pH.
- If the temperature exceeds the optimal conditions for biodegradation in the summer, then cooling strategies should be implemented for months when high temperatures prevail. Alternatively, the flow rate to the ICBs can be reduced to increase the hydraulic residence time to account for the reduced rate of biodegradation. In this case, performance will be at a decreased capacity.
- If the supply of oxygen is inadequate in the first chamber of the ICBs, the influent and recirculation piping should be modified to allow for the feed to be introduced into both the first and second chambers of each ICB unit and to permit chamber-to-chamber recirculation to better distribute the organic loading and the demand for oxygen. This will provide more flexibility to match the oxygen demand of the biomass with the available air supply.
- Although not specifically related to biotreatment performance, if the malodorous compounds released during biotreatment cannot be removed to acceptable levels, additional GAC capacity can be installed to meet this objective.

Finding 7-6. The proposed contingency options described above can maximize the likelihood that the biological treatment system will meet its target TDG removal objectives during steady-state operation.

The technical factors leading to insufficient treatment in the ICBs, along with the impacts and contingency options, are summarized in Table 7-1. Each factor is also evaluated against the performance criteria described in Chapter 6 and assigned to a performance (grade) category based on the overall risk it poses to PCAPP operations.

Finding 7-7. Any failure that leads to significant flow reduction or shutdown of the BTA for 2 weeks or more would force the cessation of agent processing owing to the limit on storage capacity for hydrolysate. This assumes the tanks with their 30-day storage and the WRS storage are partially full at the time of the failure.

As indicated in Chapter 4, adding more onsite hydrolysate storage capacity may be considered. This would provide extra time to address failures that lead to significant flow reduction or shutdown of the BTA. However, if adding hydrolysate storage capacity is considered, there should be a high degree of confidence that fixes being considered to downstream processes would be successful. Otherwise, if fixes to downstream processes are unsuccessful, adding storage capacity would only allow for storage of more hydrolysate, and the facility would still be faced with the need to consider offsite transportation. Further, considering the time it takes to process and approve permit documentation for adding storage capacity, it may not be possible to know ahead of time, prior to the decision to consider adding storage capacity, what specific issues might occur with the BTA and, consequently, whether fixes to downstream processes would be successful.

Recommendation 7-2. PCAPP should consider whether additional storage capacity might be needed and, if it is, should enter into negotiations with the CDPHE, the county, and other stakeholders to discuss options for increasing storage capacity, including adding more storage tanks or increasing containment space to accommodate contingency tanks.

FAILURE RISKS, SYSTEMIZATION, AND CONTINGENCY OPTIONS IN THE WATER RECOVERY AND BRINE REDUCTION SYSTEMS

Downstream from the ICBs, the WRS and the BRS will enable PCAPP to recover and recycle 80 percent of process water back into munitions processing and hydrolysate dilution requirements for biotreatment. The WRS serves as a holding tank where effluent from the ICBs and other process streams is collected, mixed, and stored for up to 7 days before being transferred to the BRS. No treatment or processing takes place in the WRS other than acid addition and deaeration for stripping of carbon dioxide. This section will focus on the failure risks and contingency options in the BRS.

As described in Chapter 2, the BRS is a relatively conventional evaporator and crystallizer system that separates the water from the brine, recovers the water, and, as designed, produces a filter cake, which along with the spent GAC media should be the only wastes generated by the BTA requiring offsite disposal. The unit operations in the system are conventional, but since the biotreatment process has not been tested at full scale, the composition and concentrations entering the BRS are unknown and the process has not been tested with the feed from the biotreatment units.

Technical Factors That May Lead to Insufficient Treatment

Perhaps the greatest concern for the BRS is whether the water produced by the process will meet the requirements established in the present Resource Conservation and Recovery Act (RCRA) permit by CDPHE. Chapter 4 recommends that the product water should not be considered a waste and that the only consideration is whether the water meets technical specifications for reuse. There are a number of technical factors that influence the quality of the water produced by the BRS. For example, liquid droplets carried over through the entrainment separators in the evaporator could damage the vapor compressor equipment. Droplet carryover from the crys-

TABLE 7-1 Summary of Potential Technical Factors Leading to System Failure in the Immobilized Cell Bioreactor Units, and Contingency Options

Technical Factor	Grade	Rationale for Assigned Grade	Contingency Option
TDG toxicity	0 to 1	TDG will be diluted; respirometry will identify toxicity limits. Systemization with TDG will verify treatability.	Reduce TDG loading and/or reduce flow rate to the ICBs.
Inability to control pH	1 to 2	Hydrolysate pH will be neutralized with H_2SO_4; acid generated within ICBs will be neutralized with NaOH. Systemization with TDG will verify pH control capability.	Buffer with sodium carbonate as an alternative.
Inability to control temperature	0 to 1	Steam injection ports exist for heating reactors during cold weather. No contingency as yet for cooling during hot weather, if needed.	Reduce hydrolysate throughput to accommodate slower kinetics of biodegradation during high summer temperatures.
Start-up difficulty/ acclimation	1 to 2	Systemization with TDG should facilitate smooth transition to hydrolysate treatment.	Halt start-up to address problems with hydrolysate feed.
Nutrient limitations	0 tol	Urea will be added as a source of N; DAP will be added as a source of P. Precipitation of $FePO_4$ may limit P availability; systemization with TDG will verify P availability.	DAP may be added directly to the process water recirculation line, or higher amounts of DAP may be added to the feed tank.
Oxygen limitations	0 to 1	Air will be supplied by coarse bubble diffusers to all ICB chambers to meet oxygen demand of TDG biodegradation. Systemization with TDG will verify need to redistribute influent TDG load or oxygen supply.	If oxygen demand of TDG in first chamber exceeds oxygen supply, the TDG influent can be distributed uniformly across all ICB chambers. The urea and DAP as sources of nutrients both exert an oxygen demand. Switching to nitrate and phosphate salts will eliminate this oxygen demand.
Loss/sloughing of biomass solids	1 to 2	Biomass is immobilized in ICBs, so continuous loss of biosolids should be limited; systemization with TDG surrogate will verify biomass retention and potential losses.	Increase retention time in ICBs to ensure sufficient TDG biodegradation.
Buildup of biomass solids	1 to 2	Biomass sloughing should occur naturally; systemization with TDG will help verify that solids do not build up to undesirable levels.	Increase retention time in ICBs to reduce the TDG loading rate, which will reduce the amount of biomass accumulation and sloughing.
Limited hydrolysate storage capacity	2 to 3	30-day capacity available to store hydrolysate from agent neutralization. If kinetics of biotreatment are inhibited, rate of agent neutralization can be slowed. This is not expected to be a regular occurrence but may happen intermittently.	Reduce rate of agent neutralization as needed. Construct more hydraulic buffer (storage) capacity.
Release of malodorous compounds	1 to 2	GAC adsorbers are in place to remove volatile compounds.	Install additional GAC capacity.

tallizer might have greater impact on the downstream water quality since the droplets are saturated with brine. Recommendation 3-3 in NRC 2013 states: "The concentrations of the organic compounds and suspended solids in the distillate from the PCAPP crystallizer should be carefully monitored. If they prove to be unacceptably high, consideration should be given to upgrading the de-entrainment device in the crystallizer." Another option that should be considered is to divert the condensate, as necessary, from the crystallizer to the WRS.

Operational difficulty with the GAC adsorbers is another possible cause of poor recycle water quality. NRC 2013 also points out that it may be necessary to replace the carbon in the carbon system at shorter intervals. If the pressure drop across the carbon bed is too high, the level of suspended solids in the GAC influent may be reduced by modifying the crystallizer and/or more frequent backwashing of the adsorber.

The BRS handles fluids with high chloride contents at elevated temperatures. Even though the pH is generally above 7, these fluids are highly corrosive. The material of construction selected for the BRS equipment reflects this fact, and no obviously inappropriate materials selections are noted. However, there is always the possibility that local conditions will become corrosive to the selected materials. This is especially possible in crevices or under deposits. In these locations, the pH can be much lower than in the bulk fluid, resulting in pitting or crevice corrosion. Heat transfer surfaces are especially prone to such attack.

The amount of biomass and/or organic matter that will be transferred into the BRS process is unknown. Its presence could foul heat transfer surfaces, thereby reducing the heat transfer rate and necessitating frequent cleaning. Such cleaning has been provided for by having spare heat exchangers so that fouled heat exchangers can be taken out of service for cleaning. Even so, fouling could become a problem if cleaning has to be done too frequently.

Another potential problem is that biomass carryover from the ICBs could be subjected to acid hydrolysis with the addition of sulfuric acid ahead of the deaerator, leading to the solubilization of organic matter. Excess organic matter in the BRS could result in foaming, higher organic loading to the GAC adsorbers, and less effective dewatering.

There is also a possibility that the filter cake will contain excess water, preventing it from being disposed of as a solid and thereby adding to the cost of disposal and possibly requiring some modifications to the equipment or process (e.g., addition of a chemical dewatering aid). This outcome would also modify the water balance in the overall hydrolysate treatment system and might lead to requirements for additional water to meet overall operating needs.

A possibility exists that the biomass/organic content of the filter cake will generate odors. Provisions for introducing additives into the filter cake to inhibit biological activity and eliminate potential odor issues should be considered. Such additives might include fly ash or lime prior to packaging and shipment.

Finding 7-8. Technical factors that may lead to insufficient treatment of the ICB effluent include liquid droplet carryover in the evaporator and crystallizer; failure or excessive replacement frequency of the GAC adsorbers; high chloride content, leading to corrosion; excessive biomass or organic compounds, leading to fouling, foaming, or odors; and excess liquid in the filter cake.

Recommendation 7-3. PCAPP should develop contingency plans to mitigate risk in the event that one or more of the above factors inhibit efficient BRS operations. Such plans should be in place prior to system start-up so that agent neutralization operations are not unduly delayed.

Systemization and Likelihood of Insufficient Treatment

In general, the performance of the BRS depends on the performance of the ICBs. If the biotreatment of TDG is 95 percent effective and carryover of biomass from the ICBs is minimal, then treatment in the BRS is likely to produce water with a quality sufficient for recycling back into the plant for hydrolysate dilution and munitions processing needs and a filter cake with the expected solids composition. While it is possible that factors such as fouling and corrosion will impact the water throughput for the BRS, it is unlikely that they will prevent it from functioning at all. On the other hand, if the water produced is not suitable for reuse, the system might be considered to have failed. While poor dewatering of solids in the filter cake would increase disposal costs for these solids, it is likely that the equipment or process could be modified to keep the process running (e.g., by adding sorbents to eliminate free liquid in the filter cake).

The risk reduction and mitigation effort, which involves testing surrogate hydrolysate, should greatly reduce the uncertainties identified above and should provide an early indication of the nature and extent of any problems that might arise. Important data to be gained from these tests include the following:

- Composition and concentrations of components entering BRS.
- Total organic carbon (TOC), total suspended solids (TSS), and total dissolved solids (TDS) in crystallizer distillate.
- TOC, TSS, and TDS to GAC adsorbers.
- GAC adsorber effluent quality.
- Pressure drop in carbon adsorbers.
- Filter cake water content and composition.
- Equipment records, especially heat exchanger, fouling, and required cleaning schedule.

Finding 7-9. Data collected during risk reduction and mitigation activities can be used to assess how closely BRS operation matches its performance specifications. This information will allow a determination of the likelihood and significance of failure of the BRS or components therein over time.

Recommendation 7-4. PCAPP personnel should identify and monitor critical process data during risk reduction and mitigation activities. These data could include composition and concentrations of solutions entering the brine reduction system and leaving the granular activated carbon (GAC) adsorber, total organic carbon, total dissolved solids, and total suspended solids in process streams, pressure drop across the GAC, filter cake water content, and equipment fouling.

Finding 7-10. If, for some unforeseen reason, the BRS does not produce water of acceptable quality for reuse, there will be no way to recycle the water within the plant at PCAPP. This failure will place much greater strain on the aquifer from which PCAPP withdraws water. Moreover, the effluent from the ICBs will have to be shipped offsite for treatment and disposal. Because the hydrolysate is diluted eightfold prior to entering the ICBs, the liquid volume leaving the ICBs will be eight times that of the hydrolysate.

Recommendation 7-5. If the brine reduction system fails and a decision is made to ship offsite, it would be more prudent to ship undiluted hydrolysate rather than continuing to operate the immobilized cell bioreactors and ship the effluent offsite. This action would minimize the total volume of material that needs to be shipped offsite and, although it would still place additional strain on the freshwater needed from the aquifer, it would minimize that quantity.

As discussed in Chapter 4, under the current RCRA permit, process water from the BRS can be recycled within the plant only if it meets Colorado Drinking Water Standards for all RCRA constituents (Appendix VIII of Section 261) or if, for a specific application and with CDPHE approval, it meets the limitations for all RCRA constituents. Based on the recommendations in Chapter 4, the only consideration should be whether the BRS process water meets technical specifications for reuse within the plant. Should a modification of the current permit requirements for water recycling be rejected by CDPHE, there is a moderate probability that BRS water treatment requirements cannot be met consistently, even though the water is still of adequate quality to be reused within the plant. Therefore, it would be prudent to apply now for permit modifications or other actions, as recommended in Chapter 4.[2]

Impacts If WRS or BRS Underperform or Do Not Perform

One of the main purposes[3] of the WRS and BRS is to reclaim water that can be recycled back into the munitions processing and agent neutralization operations and for dilution of the agent hydrolysate. If the ICBs perform effectively but the downstream WRS and BRS underperform, there would be a significant impact on the ability to reuse the process water. If the recycle water does not meet drinking water standards but is still of adequate quality for the process applications noted, the permit could be modified to satisfy an alternative water standard—for example, the wastewater discharge standard.

If the BRS water that is recovered for reuse comes close to meeting the product water quality requirements but is not completely suitable for recycle, then two solutions could be evaluated: (1) determine the source of the undesirable contamination and selectively remove that water stream from the WRS/BRS water feed system or (2) install additional water treatment to allow the feed water to meet process water requirements. Either case will require permit modification as well as the identification of an appropriate management system for the water.

[2] The reader is referred to Chapter 4 for an in-depth discussion on permitting related to the quality of recycled product water, along with recommendations on actions PCAPP can take to reduce the risk of failure.

[3] Another important purpose of the WRS/BRS is to produce a filter cake with no free liquids that may then be disposed of in compliance with regulatory requirements without further treatment.

Contingency Options

A variety of steps can be taken in the event that the BRS does not perform as expected. Some of these steps involve plant modifications; others are more regulatory in nature. A number of potential causes of underperformance (i.e., technical factors) will be tested during systemization. If issues are identified, then PCAPP can prepare and implement contingency plans as needed, before full-scale operations begin. While many potential causes of underperformance will be tested during risk reduction and mitigation efforts and during systemization in general, the full extent of the impact may not be realized until full-scale munitions processing begins:

- If there is liquid droplet carryover through the entrainment separator, the crystallizer de-entrainment device could be upgraded. This would incur additional infrastructure costs that would, however, probably outweigh the costs of the damage that could be caused by such carryover.
- If the crystallizer condensate is not of adequate quality, it could be separated and sent back to the WRS for reprocessing. This would reduce the throughput of the BRS, thereby reducing the product water available for recycle. However, as long as the storage capacity of the WRS is not exceeded, the improvement in performance might be a favorable trade-off for the decreased capacity.
- If there is excessive biomass carryover from the ICBs to the WRS and BRS, it may be necessary to install a clarifier after the ICBs to remove the biomass and other TSS ahead of the WRS. Depending on the hydraulic loading rate of the clarifier, this addition may slow process throughput but improve performance of the BRS.
- If there is excess liquid in the filter cake and no options exist for modifying the BRS to remove the excess liquid (e.g., drying agent additives such as $CaCO_3$ are ineffective), alternative arrangements will be needed for offsite shipment and disposal.
- If the solid residue in the filter cake has unacceptable odors, additives could be applied to control the biological activity responsible for the odors.
- Regular monitoring and maintenance of the systems is critical to successful operations. However, overly demanding maintenance requirements (e.g., too-frequent replacement of GAC or cleaning of heat exchangers due to fouling) can be costly and can cause schedule delays. These issues may be best addressed by incorporating more system redundancy or by reducing system throughput. The impacts of these operational changes must be weighed against the impact of excessively frequent maintenance.

TABLE 7-2 Summary of Potential Technical Factors Leading to System Failure in the Brine Reduction System, and Contingency Options

Technical Factors	Grade	Rationale for Assigned Grade	Contingency Options
BRS product water quality is acceptable for reuse but does not meet permit requirements.	2 to 3	Product water may not consistently meet RCRA permit requirements.	Initiate obtaining permit modification to adjust the BRS water treatment requirements.
Liquid droplet carryover through the entrainment separators.	1 to 2	Liquid droplets in the evaporator could damage compressor. In the crystallizer, they could reduce recycle water quality.	Upgrade de-entrainment devices.
Poor performance of GAC adsorbers.	0 to 1	High TOC, suspended solids, or microbial growth on GAC lead to need for backwash due to the large pressure drop across adsorbers or frequent replacement of GAC.	Install additional GAC capacity, standby GAC adsorbers.
Corrosion, especially on heat transfer surfaces.	1 to 2	High chloride content and high temperatures could lead to corrosion in crevices and under deposits.	Aggressive corrosion monitoring and deposit removal are required to avoid unexpected failures.
Foaming and fouling, especially on heat exchangers.	1 to 2	Biomass carryover from ICB can lead to acid hydrolysis of biomass and solubilization of organic matter; currently unknown how much biomass will be carried over.	Install clarifier after ICBs to remove biomass and other TSS; monitor antiscalant effectiveness and fouling tendencies; add antifoaming agents; increase cleaning frequency.
Filter cake with biological activity and organic material.	0 to 1	Biological activity may lead to unacceptable odors.	Include additives (e.g., fly ash or lime) in filter cake to inhibit biological activity.
Liquid content of filter cake is too high.	0 tol	Treatment in BRS does not yield a solid product owing to high organic content; drying agents used as additives are insufficient to dry the cake.	Find an alternate TSDF that will accept waste.
pH control.	0 to 1	Low pH is required for CO_2 stripping in the WRS, and a neutral pH is required for the BRS to minimize the corrosion potential.	Control pH in the WRS with sulfuric acid and control pH in the BRS with NaOH.

NOTE: TSDF, treatment, storage, and disposal facility.

The technical factors leading to insufficient treatment in the WRS and BRS are summarized in Table 7-2, along with the impacts and contingency options. Each is also evaluated against the performance criteria described in Chapter 6 in order to assign a grade to the overall risk it poses to PCAPP operations.

OFFSITE SHIPMENT AS A CONTINGENCY OPTION

Up to this point, Chapter 7 has focused on operational strategies to maintain onsite capabilities for treatment of hydrolysate in the event that the BTA does not meet selected performance goals. However, if underperformance of the overall system cannot be remedied through the strategies described above, then offsite shipment of hydrolysate may become a better alternative. While the offsite shipment and disposal option is viewed as a last resort, it is prudent to plan and prepare in advance for such a scenario in the event it is needed in order to avoid further delays in agent neutralization operations. This section discusses the downstream impacts of a decision to ship hydrolysate offsite and the steps that would need to be taken to make this operational change.

Downstream impacts of, and actions needed to implement, the offsite shipment option are described here in terms of "plant, paper, and people." The committee and PCAPP leadership discussed the downstream impacts of offsite shipment of hydrolysate (see Lecakes [2014] for a complete list of the topics discussed). Many factors highlighted in this section were addressed at that meeting. Although the committee views it as unlikely that offsite shipment of hydrolysate will be needed, PCAPP can initiate preparations for such shipment now, prior to live agent operations, to expedite the implementation of this option should it be needed.

Plant

Should the decision be made to ship hydrolysate offsite, additional infrastructure would be needed to efficiently and effectively transfer the hydrolysate for shipment. Shipment could occur by truck, rail, or some combination of the two. Necessary infrastructure might include additional piping and storage tanks, leak and odor containment, MINICAMS[4] for agent monitoring, waste loading areas, loading docks, new rail-/roadways onsite, and extra traffic controls. It is of note that rail infrastructure currently exists at PCD but not yet within the PCAPP facility.

An important impact of the decision not to use the BTA is that the plant will be unable to recycle process water. Although there will be water savings in the overall process since the need for a 1:7 dilution for ICB operations will be eliminated, a significant amount of water will still be consumed during munitions processing and agent neutralization. PCAPP might need to obtain approvals for increasing amounts taken from groundwater or might be required to identify supplementary water sources to offset the shortfall. The Pueblo area is under water stress, so drawing from local water resources may put undue burden on the surrounding community. Furthermore, new infrastructure at PCD may be needed to supply additional water to PCAPP. A contingency plan for such additional water supply should be developed ahead of any offsite shipment decision.

Finally, consideration must be given to how the hydrolysate will be treated elsewhere. An appropriate TSDF should be identified ahead of any offsite decision, as discussed in Chapter 4. Once hydrolysate is delivered to its destination, one disposal option is deep-well injection; treatment options include biotreatment and/or incineration. Although significant investments in labor, time, and money have been devoted to the BTA, there are some benefits of not operating this portion of the plant. These include a reduction in odor and emissions; elimination of solid waste streams (e.g., BRS filter cake, spent carbon from the GAC adsorbers); a reduction in utility loading (e.g., heat tracing, cooling water, and steam); and a significant reduction in maintenance and manpower needs. Existing plans for closure of the BTA could be initiated if the offsite shipment scenario is implemented.

Paper

Should the decision be made to ship hydrolysate offsite, a significant amount of paperwork for permitting, operating procedures, and contracts must be in place before actions can be taken to implement the decision. Chapter 4 discusses permit requirements applying to changes in PCAPP operations and offsite shipment of hydrolysate in detail. Briefly, offsite shipment will impact the facility's RCRA permit, may impact the NEPA requirements, could impact the Pueblo County Certificate of Designation, and will also impact adherence to OPCW treaty requirements. Furthermore, PCAPP will need to identify and contract with licensed hauler(s) and TSDF(s) and coordinate shipment to the final disposal site (see Chapter 5). Changes to the PCAPP facility's Site Safety Submission Document and Facility Construction Certification would be required, along with revisions, cancellations, and the adoption of new standard operating procedures. "Paper" is an area where preplanning for this worst-case contingency option might be very beneficial.

People

As indicated earlier, there are numerous stakeholders in the PCAPP project. A major change in operations, such as a shift to offsite shipment of hydrolysate rather than use of the BTA, would have a major impact across all stakeholder groups. In particular, shutdown of the BTA would likely result in PCAPP staff reductions or reassignments, and delays in implementation of offsite shipment could result in further loss of staff from the facility as a whole. The surrounding community, represented by the Colorado CAC, would also be impacted by this decision. Transparent decision making and frequent communications with the Pueblo community may contribute to their support and cooperation in the event that offsite shipment is needed, as discussed in Chapter 3.

Recently, the Colorado CAC briefly presented its perspective on the risks of offsite shipment of hydrolysate (Kornelly, 2014, reprinted in Appendix B). These risks include permitting changes, costs, public perception, and operational logistics. The Colorado CAC also identified a list of things that "could go wrong" in the BTA. All the technical factors raised by the Colorado CAC are considered in this chapter. It is worth noting that the Colorado CAC has high confidence that these technical factors can be addressed by PCAPP, primarily through rigorous maintenance. Overall, PCAPP is well positioned for successful operations.

As described earlier in this chapter, there are many actions that can be taken should the BTA not operate as expected. Any decision on the implementation of a contingency option should, as discussed in Chapter 6, be based on an established decision framework that takes into consideration a variety of criteria—such as effectiveness, efficiency, cost, and regulatory factors. The committee acknowledges that there are many uncertainties surrounding the start-up and performance of each separate component within the BTA and their ability to work together to treat the hydrolysate, and one or more contingency options may have to be resorted to. Each decision may have to consider a continuum of options, from operational tweaks to improve performance (e.g., changing chemicals to maintain pH levels) to more long-term operational changes (e.g., longer retention times) and infrastructure changes (e.g., installing a clarifier) to accommodate performance issues, to interim actions while other contingency options are being implemented (e.g., constructing and employing additional hydrolysate storage capacity), and, finally, to instituting offsite shipment of hydrolysate.

Such options may constitute a temporary change (e.g., a matter of days, weeks, or a month or two until a remedy

[4] MINICAMS are automated near-real-time monitoring systems that have been used in chemical agent disposal facilities since 1990 to monitor for chemical agents.

can be implemented) or a permanent one, and decision making using the performance criteria and decision framework should consider both options. For instance, if the change is only temporary, there could be additional actions or costs necessary to restart operations in the BTA. Such decisions might take into account more than just technological or cost criteria. For instance, temporary shutdowns could result in perceived uncertainties leading to the loss of trained PCAPP staff and loss of confidence on the part of the community.

The committee believes that the optimum outcome is that the existing BTA operate without the need to implement the offsite option. The committee sees offsite shipment of hydrolysate as the last resort, the final option on the continuum. However, if offsite shipment of hydrolysate is chosen, one very crucial decision that will need to be made is whether the offsite shipment will be temporary or permanent. The committee acknowledges the possibility that once the decision to implement offsite hydrolysate shipment is made, it may be necessary to make that process permanent due to cost, the need for stability, or other considerations. The committee also acknowledges that the fix or set of fixes needed for the BTA may take only a few days, or weeks, or even a month or two, and that it may be possible, after some delay, to start up the process again and continue with onsite hydrolysate treatment.

Implementing offsite transport of hydrolysate will affect plant, paper, and people, as discussed above. Physical changes to the plant, changes in permit documentation and standard operating procedures, retraining and possible reassignment or even furlough or layoff of staff would be needed, to name just a few examples. The effort to implement offsite transport will be considerable. If it is implemented as a temporary fix, with the intent of restarting the BTA processes, the effort to switch back to the BTA would also be considerable. Depending on the length of the delay and on whether staff furloughs or layoffs have occurred, original staff may no longer be available. Besides which, if the BTA processes are restarted, there is no guarantee that the fix will even work, and PCAPP may need to restart offsite shipment again. Still, the committee believes that there may be circumstances under which restarting the BTA processes, after some delay, may be feasible. The committee discussed at length whether a change to offsite shipment could be temporary, or whether this change should be permanent. It acknowledged that at this time it is impossible to predict the exact circumstances leading to severe underperformance of the BTA. The committee concluded, therefore, that it would make no specific recommendation concerning the exact nature, extent, or permanence of any option, including the permanent, offsite shipment of hydrolysate. The decision must be based on the application of an established decision framework and appropriate consultation.

Finding 7-11. The decision to ship hydrolysate offsite can result in many types of downstream impacts, including the addition of new infrastructure for shipment, modifications to permits and new operating procedures, and PCAPP staff reductions or reassignments.

Recommendation 7-6. In the event that offsite shipment is the only viable option for PCAPP to meet its mission goals, this contingency option should be implemented as efficiently as possible to reduce the downstream impacts that could result in a significant delay in munition processing.

Recommendation 7-7. To preserve the ability to ship hydrolysate offsite for treatment in the event that offsite shipment is found to be the only viable option, steps should be taken as soon as possible. Examples of such steps include initiating permit modifications; drafting alternative standard operating procedures; preparing transportation risk documentation; designing process safety controls, spill containment, and fall protection for hydrolysate loading facilities; and communicating with stakeholders about if and when this option would be utilized, including how the stakeholders would be involved in the decision process.

REFERENCES

Guzzetti, Don. "Biotreatment Area Risk Reduction and Mitigation," PCAPP presentation to the committee on July 29, 2014.

Kornelly, Irene. "PCAPP Bio-Treatment Area Concerns of the Colorado Chemical Demilitarization Citizens' Advisory Commission (CO CAC)." August 29, 2014.

Lecakes, George. "Downstream Impacts to Plant Operations If Offsite Shipment Is Required," PCAPP chief scientist, discussion facilitated between committee and PCAPP leadership on July 29, 2014.

Metcalf and Eddy/AECOM. 2014. *Wastewater Engineering: Treatment and Resource Recovery,* 5th Edition. New York, N.Y.: McGraw-Hill Education.

NRC (National Research Council). 2013. *Review of Biotreatment, Water Recovery, and Brine Reduction Systems for the Pueblo Chemical Agent Destruction Pilot Plant.* Washington, D.C.: The National Academies Press.

PCAPP (Pueblo Chemical Agent Destruction Pilot Plant). 2014. *White Paper Bio-Treatment Area Risk Reduction and Mitigation.* 24852-30H-BTA-V0001. Pueblo, Colo.: Pueblo Chemical Agent Destruction Pilot Plant.

SBR Technologies, Inc. 1996. *Biodegradation of HD Hydrolysate in Sequencing Batch Reactors.* South Bend, Ind.

Appendixes

A

Statement of Task

The project is to be undertaken in two reports: a first report, on the Pueblo Chemical Agent Destruction Pilot Plant (PCAPP), to be delivered 9 months from start of contract, and a second report, on the Blue Grass Chemical Agent Destruction Pilot Plant (BGCAPP), to be delivered at the end of the 15-month contract period.

The National Research Council will establish an ad hoc committee to consider the following study objectives for the PCAPP first report:

- Review the criteria for successfully treating the hydrolysate;
- Determine technical factors, decision points, and contingency options if the biotreatment and brine reduction systems either underperform or do not perform at all, thereby impacting destruction operations;
- Identify systemization data that should factor into the criteria/decision process;
- Identify any possible plant modifications that would allow continued use of the onsite processing, e.g., use of additional buffer capacity, increased dilution, shipment of brine reduction system brines, and shipment of excess amounts of hydrolysate beyond onsite capacity;
- Review downstream impacts to plant operations if offsite shipment is required;
- Identify any permitting requirements for offsite hydrolysate shipment and treatment options, and evaluate transportation risks that could be expected;
- Consider stakeholder interests and solicit stakeholder input.

B

Public Interest and Input Documents

Public Involvement in NRC Hydrolysate Study

Dr. Judith A. Bradbury, Member

Committee on Review of Criteria for Successful Treatment of Secondary Waste and Hydrolysate at the Pueblo, CO and Blue Grass, KY, Chemical Agent Destruction Pilot Plants

Public Involvement in NRC Hydrolysate Study

- Three-pronged approach proposed to the CAC for providing opportunities for public input:
  - CAC meeting
  - Dedicated NRC email address for ongoing input
  - Webinar combining written and verbal input, to permit clarification and further discussion of outstanding issues (early October)

Please share your perspectives on shipping hydrolysate from PCAPP, by email to:

Comments_for_NRC_Hydrolysate_Committee@nas.edu

Hydrolysate Committee
500 Fifth Street, NW, Rm 934
Washington, DC 20001
Fax: 202 334 2620

THE NATIONAL ACADEMIES

Division on Engineering and Physical Sciences
Board on Army Science and Technology

Public Involvement in NRC Hydrolysate Study

- Public involvement is a critical study component: two experienced committee members assigned to the task
- Goal in data gathering is to listen to and consider the range of views of the CAC and affected public

Public Involvement in NRC Hydrolysate Study

- Schedule for providing public input is driven by the schedule for report completion
- Comments and suggestions for enhancing public input (within schedule constraints) are welcome

Exhibit 1: Committee Approach to Enhancing Public Input: Presentation to the Colorado Citizens' Advisory Commission, July 30, 2014.

(continues)

Public Involvement in NRC Hydrolysate Study

- CAC members are invited to lunch and dinner and to attend the NRC open sessions when contractor presentations will be given
- Scheduled open sessions are at:
  - Pueblo Convention Center
  - July 29, 1:00 – 5 p.m.
  - July 30, 7:45 a.m. – 4:30 p.m.

HANDOUT CARDS

Please share your perspectives on shipping hydrolysate from PCAPP, by email to:

Comments_for_NRC_Hydrolysate_Committee@nas.edu

THE NATIONAL ACADEMIES
Advisers to the Nation on Science, Engineering, and Medicine

Hydrolysate Committee
Board on Army Science and Technology
500 Fifth Street, NW, Rm 934
Washington, DC 20001
Fax: 202 334 2620

Exhibit 1: Continued.

PUEBLO CHEMICAL DEPOT

Chemical weapons waste shipment eyed

National Research Council wants to gauge local attitudes on proposal

BY CHRIS WOODKA
THE PUEBLO CHIEFTAIN

The National Research Council is seeking the views of Pueblo residents on the potential off-site shipment of hydrolysate from the Pueblo Chemical Agent Destruction Pilot Plant after the chemical agent is destroyed.

PCAPP is a facility designed to destroy the stockpile of 780,000 chemical weapons containing 2,600 tons of mustard agent currently stored at the Army's Pueblo Chemical Depot, 15 miles east of Pueblo. Destruction will begin soon and is expected to be complete by the end of 2017.

Hydrolysate is the waste product that remains following the chemical destruction by hydrolysis of the chemical agent.

The NRC is responsible for determining conditions under which off-site transport of the hydrolysate may be necessary to ensure continued destruction of the stockpile.

"Part of our tasking is to consider stakeholder interests and solicit stakeholder input," said committee chair Dr. Robert A. Beaudet, an emeritus professor of chemistry at the University of Southern California. "We want to understand the perspectives of the people who live and work in Pueblo."

> "We want to understand the perspectives of the people who live and work in Pueblo.
>
> *DR. ROBERT A. BEAUDET, EMERITUS PROFESSOR OF CHEMISTRY AT THE UNIVERSITY OF SOUTHERN CALIFORNIA*

The committee also is planning to hold a webinar to describe its efforts and take comments and suggestions if sufficient interest is expressed by members of the Pueblo community. Comments or inquiries should be sent to: Comments_for_NRC_Hydrolysate_Committee@nas.edu.

cwoodka@chieftain.com

Exhibit 2: Article in *The Pueblo Chieftain*, September 5, 2014, Publicizing Dedicated E-mail Address for Comments from Stakeholders.

Remarks Made to the
National Research Council
Hydrolysate Committee
By
Irene Kornelly
Chair, Colorado Citizens Advisory Commission

July 29, 2014

The Pueblo community has been an active participant in the destruction of the chemical weapons stored at the Pueblo Chemical Depot from the very beginning of the program. When the Department of the Army originally proposed that the weapons stored at the Depot would be destroyed by incineration, a vocal portion of the community strongly objected to this method. With vigorous lobbying of the Colorado and Kentucky Congressional delegations, the community was pleased to see the passage of legislation that led to the establishment of the Assembled Chemical Weapons Assessment program which became the Assembled Chemical Weapons Alternatives (ACWA) program. Members of the community were active participants in the ACWA Dialogue program and the community followed with great interest as a technology was chosen for Pueblo. On the evening when the CAC voted in favor of the Neutralization/Bio-treatment technology, one of the members gave an impassioned speech and said "it is time to vote for a technology that represents the future and not one that represents the past." Thus with Department of Defense concurrence, the ACWA program pursued the "technology of the future."

Originally, the Neutralization/Bio-treatment technology was intended to include not only the neutralization of the agent, but also the energetics and all hydrolysate was to be treated on-site. When DoD determined that the original design was too expensive, all work on the facility came to a halt and the facility was re-designed. With great reluctance the CAC and other stakeholders agreed to off-site shipment of uncontaminated energetics. Work slowly resumed and we are at this moment in time today.

The question of hydrolysate shipment has been under discussion by the ACWA communities – Pueblo and Blue Grass - <u>at least since 2003</u>. <u>Numerous studies</u> have

Exhibit 3: Remarks by Chair of Colorado Citizens' Advisory Commission to NRC Hydrolysate Committee, July 29, 2014.

(continues)

been conducted costing an unknown, but presumably, large amount of money. Not to mention the time, attention and effort given to this issue by all parties involved. The most recent was a November 2008 report by Noblis and a November 2008 report by the NRC which looked at all secondary wastes. The November 2008 report indicated that the greatest benefit for both cost and schedule would be realized if the ACWA program acted immediately and that program benefits decreased significantly if actions were delayed. The reports only took the timeline to 2013, but by that time all cost savings were effectively eliminated. A Department of Defense Acquisition Memorandum was issued that clearly indicated that off-site shipment of hydrolysate was no longer under consideration. Now we are back again to look at hydrolysate shipment in the event that the Bio-Treatment Area at the Pueblo Chemical Agent-Destruction Pilot Plant (PCAPP) does not operate successfully. We in the Pueblo community are weary and growing increasingly impatient with continually having to re-address this issue.

The concerns that have been raised about off-site shipment of hydrolysate essentially have remained the same over the years. Of primary consideration must be the issue of dumping our wastes on someone else. While no one is happy that the chemical weapons are stored at the Depot, the issue is a fact and this fact must be dealt with in a responsible manner. Citizens of Pueblo and Colorado, in general, disapprove when someone dumps their trash into a neighbor's yard, and the reaction is the same when it is suggested that we should dump these industrial wastes (hydrolysate) in another community. This concept is simply wrong. Other concerns that must be addressed are:

1. Identification of transportation routes, both in Colorado and through other states to a Treatment Storage and/or Disposal Facility (TSDF).

2. Identification of the TSDF to be used. The 2008 Noblis report identified three types of TSDF facilities that could potentially accept hydrolysate. These types of facilities included biotreatment facilities, deep-well injection and incinerators. Each of these methodologies has their own pluses and minuses. Deep-well injection and incineration are the most problematic to Pueblo.

3. Assessment of the potential political reaction from other communities affected by plans to move hydrolysate associated with chemical weapons into or through

Exhibit 3: Continued.

their communities. While Pueblo has a very active Citizen Advisory Commission and an equally active Outreach Office capable of providing information and education to the community, other communities do not have that advantage. Given past experiences, it seems clear that acceptance issues could pose real threats to completing the project in Pueblo in a timely fashion. The community concerns and the delays associated with properly addressing those concerns have the potential to reach beyond the control of the parties currently engaged in the process (DOD, ACWA, the State of Colorado and the Pueblo community), who have the most to gain from transportation. Host communities and communities along transportation routes have little to gain and potentially more to gain from delaying the process if they have concerns (legitimate or not) about the hydrolysate from PCAPP entering or passing through their communities. What are the contingency plans in shipment of hydrolysate to a particular TSDF becomes impossible?

4. The concern about the efficiency and thoroughness of the coordination of all notification, monitoring and management of the hydrolysate shipments, including all of the various federal, state and local agencies that would need to be involved, or should be involved. This issue not only includes the State of Colorado, but all states throughout any proposed transportation routes and the ultimate destination state.

5. The high potential that many communities along the transportation route may want to use this opportunity to enhance their local hazardous waste management and emergency response programs at the expense of this program.

6. The cost of clean-up of a hazardous waste spill as a result of a transportation accident. The costs should be assessed and included as a potential cost to the project. While we have been assured that the shipment of hydrolysis is no more dangerous that shipping "Drano," I for one do not want spilled Drano in streams or on road surfaces or ditches. It is a dangerous liquid and on the side of each can there are extensive warnings about the toxicity of the contents.

Exhibit 3: Continued.

7. The compliance record of the facilities to which the agent hydrolysate might be sent and the financial assurance that the TSDF can not only safely destroy the hydrolysate, but will do so in a timely manner. The compliance record should include OSHA compliance as well as environmental compliance, including Environmental Justice.

8. The potential that the change in the hydrolysate treatment methodology would trigger the need for additional NEPA review as well as changes to the RCRA permit and the Certificate of Designation issued by Pueblo County. The result could be a slow down or interruption in chemical weapons destruction. These delays cost money and pushes out even further ultimate compliance with the Chemical Weapons Convention.

Since the issue under consideration is specifically the potential shipment of hydrolysate should the Bio-Treatment system fail, there are additional concerns that should be addressed.

1. For years the citizens of Pueblo have been assured that the operations of the Bio-Treatment system is no more complicated than a bio-treatment unit at any municipal waste water treatment system throughout the company. In addition Puebloans have been assured that both Bechtel and ACWA have hired the best engineers in the country to ensure that the system will work properly. Now we are asked to consider that the system may not work in accordance with best operating procedures and that we must consider criteria in the event of system failure. These two statements do not compute.

2. One of the first things people in the community asked me when the formation of this committee was announced was "if the ACWA program is rescinding their agreement to treat the hydrolysis on site?" The same question was raised by members of the CAC. Trust between the community and PCAPP is difficult to measure when it is there, but easy to measure when it is not there. A decision to ship hydrolysate would result in a loss of trust between the community and PCAPP.

Exhibit 3: Continued.

3. Education and communication are a crucial part of continuing trust and understanding between the community and PCAPP. I am grateful to Mr. Whyne for providing the CAC and the community with a heads-up about the formation of this committee. His assurance that the committee was formed to provide a set of criteria to determine successful treatment of the bio-treatment system and not as a committee to intentionally kill the bio-treatment system was a big factor in community acceptance. While there is still skepticism, the very fact that you are in Pueblo and willing to attend the Permitting Work Group and CAC meetings and allow the public to attend your deliberations goes a long way to eliminate skepticism and distrust.

4. If hydrolysate is shipped to a TSDF, we must consider the loss of water from the facility. The recycling of water was a major component in the selection of a technology for Pueblo where water is a precious commodity. It was previously estimated that shipment of the agent hydrolysate would result in a loss of about 17 acre feet of water. The shipment of hydrolysate would be an expense to the ACWA program because additional costs for water would have to be included. At one time several years ago when it was suggested that the empty tanks could be filled with water on their return trip from New Jersey or Texas, the comment was met with laughter and derision and officials were told in no uncertain terms that water from New Jersey was unacceptable.

5. Lastly, the potential loss of jobs and economic opportunities for the Pueblo community must be addressed if transportation of the hydrolysate is considered a viable option. The citizens of Pueblo have stepped forward for years, allowing these chemical weapons to be stored, handled and disposed of in their backyard. They have directly participated in the construction of the PCAPP facility. That construction project includes the bio-treatment plant. We have built it and we should not be robbed of the economic benefit of operating the plant to its fullest capacity.

Some recommendations that the CAC would ask that the NRC Committee consider when determining the shipment of hydrolysate are:

Exhibit 3: Continued.

1. The CAC and stakeholders must be included in establishing the criteria and specific decision points that would be used to determine that hydrolysate shipment is the only remaining option.

2. The CAC and stakeholders must be included in the decision making process should the question of hydrolysate shipment ever need to be addressed.

3. Issues of cost and safety must be considered as well as the impact to local and state law enforcement that will regulate shipment issues.

4. Shipment of the hydrolysate must not be the first and only remedy but the remedy of last resort. Careful analysis should be done in order to determine if there are fixes to the operation. Is one portion of the system inoperable while the rest is operable? Is the inoperable portion so crucial that the remainder of the system cannot be used? All other options must be thoroughly explored and found to be not feasible before the transportation option is ever given serious consideration.

5. Continuous communication must transpire with the community and CDPHE to avoid misunderstandings and obstacles to hydrolysate shipment and facilitate permitting changes that must be approved by the State and County.

On a personal note, I would like to thank you for the opportunity to provide you with the concerns of the CAC. We are a group of citizens that has as its one and only goal the destruction of the chemical weapons stored at the Pueblo Chemical Depot in a safe and environmentally sound manner. Our neighbors and friends work at PCAPP and live in the community and it is our desire to see them come home every evening.

Exhibit 3: Continued.

Colorado Chemical Demilitarization Citizens' Advisory Commission

Irene Kornelly, Chair
1602 Clemson Drive
Colorado Springs, CO 80909
Home 719-591-5157
Cell 719-330-2359
Fax: 719-591-1305
Email: ikornelly@pcisys.net

Members:
Col Jeff Chostner, USAF (Ret.)
Terry Hart, Vice Chair
Kenneth Griffin
John Norton
Joe Schieffelin
Aaron Serna
Ross Vincent

PCAPP Bio-Treatment Area
Concerns of the Colorado Chemical Demilitarization Citizens' Advisory Commission (CO CAC)
August 29, 2014

The Bio-Treatment Area (BTA) is the final step in the chemical weapons destruction process at the Pueblo Chemical Agent-Destruction Pilot Plant (PCAPP). The BTA components are the Bio-Treatment System (BTS), the Water Recovery System (WRS) and the Brine Reduction System (BRS). The key to a successful BTA is high-reliability of the processes and the ability of the BTA to meet the output of the neutralization process. Potential failures, risks and mitigation measures to be taken will be a matter of overcoming the limitations of the BTA. In the opinion of the CO CAC, a catastrophic failure of the BTA requiring shipment of the hydrolysate off-site would be a total failure of the biomass or the failure to properly reduce the thiodiglycol (TDG) to 86% or greater. ACWA and PCAPP personnel, however, should be able to mitigate both of these issues.

While it may seem to be a simple matter of shutting down the BTA should one or all parts of the systems fail, there are also risks to the decision to ship the hydrolysate to an unknown destination. These risks come with a monetary cost as well as a loss of schedule. The cost savings that were identified in previous hydrolysate shipment studies are no longer valid as changes had to be made by the end of 2013. The real cost savings identified in these studies were in the reduction in construction costs which are now complete.

Briefly listed, the risks to shipment of hydrolysate are:

- New NEPA documentation – a minimum of an Environmental Assessment would have to be written which could take 6 months or more. Any attempt to push an environmental document without thorough study would cause even a greater delay.
- Class 2 modification of the RCRA permit – another 6 month delay.
- Changes to the Certificate of Designation (CD) by Pueblo County.

Exhibit 4: Follow-up Statement of Concerns from the Colorado Citizens' Advisory Commission, September 29, 2014.

(continues)

PCAPP Bio-Treatment Area
Concerns of the CO CAC Page 2

It is conceivable that some portions of these three tasks could be pursued concurrently, but it is also likely that they will be seen and treated as interdependent at some points in the process, so final completion of one or more of them could be delayed. In addition both the County and CDPHE could grant a Temporary Authority (TA) to PCAPP to begin construction of the loading dock and piping modifications prior to the RCRA approval. However, it is unknown if CDPHE and the County would grant the TA prior to the completion of the NEPA process.

In addition to the changes in NEPA and permitting there are other issues to be considered:

- Modifications would have to be made to the Air Permit.
- Negotiations with the OPCW would have to occur to ensure that the thiodiglycol (TDG) destruction at another facility was tracked as a part of the Chemical Weapons Treaty.
- Costs for the construction of the dock, piping and road access to the dock would have to be included.
- Extensive coordination between federal, state and local agencies would have to occur to ensure that transportation of the hydrolysate was done according to all laws and regulations. The distance between PCAPP and the receiving facility would determine the complexity of this task.
- Since no water would be recycled due to the shipment of the hydrolysate, water costs for augmentation would have to be negotiated with the Pueblo Board of Water Works.
- Unless an extensive educational campaign was begun, public sentiment in Pueblo would, in general, be opposed to the shipment of hydrolysate. The neutralization/bio-treatment process was presented to Pueblo as a total package. The discussions within the community that have gone on for many years have all reached the same conclusion – do not ship the hydrolysate. A change to this decision would be received with skepticism.

What could go wrong in the BTA and can these potential failures be mitigated? The following is a list of potential concerns in the BTA process, all of which should be able to be mitigated. Bechtel and ACWA are already taking steps to check out each of these concerns with early testing. This knowledge will assist in determining which mitigation measures will have to be taken and what additional maintenance will have to be performed in order to make the BTA process a success.

1. Recycled water does not meet the drinking water standards as established by CDPHE. Can this issue be re-negotiated with CDPHE to meet a lower standard, but still have the water be re-usable in the hydrolysate process? The answer is probably "yes" and very preliminary discussions have already begun with CDPHE.

2. Air emissions at the BTA do not meet the MPHRA model. Mitigation measures can be taken to ensure that the MPHRA model can be met. The maintenance of the charcoal filters and the use of additional charcoal will add costs to the process.

3. Odors from the BTA are unacceptable and go beyond the boundaries of the Depot. While odor issues are a concern in any bio-treatment area, the extent of the odors can be mitigated so that they are acceptable beyond the boundaries of the Depot.

Exhibit 4: Continued.

PCAPP Bio-Treatment Area
Concerns of the CO CAC Page 3

4. The WRS does not meet the current plan to recover 86% to 95%. In this case costs for water would increase, but this is not a deal breaker. Any recycling of water is an asset.

5. Equipment within the BTA could fail. Equipment failure can be reduced through a more rigorous maintenance schedule.

6. Corrosion could be a factor. Equipment failure can be reduced through a more rigorous maintenance schedule.

7. Leaks within the BTA. Equipment failure can be reduced through a more rigorous maintenance schedule.

8. pH cannot be controlled. This is an issue of chemistry and ensuring that pH can be controlled.

9. Temperature extremes impact the biomass. Testing is already underway to determine how the extremes in outside temperature coupled with the temperatures created by the biomass and equipment can be mitigated.

10. Inadequate TDG removal which must be at least 86% according to the CWC. If this occurs, this is a potential reason for system failure.

11. Biomass seed does not acclimate to the hydrolysate feed. This is almost a subset of #10 and could also be a potential reason for BTA failure.

12. Solids accumulation in the ICBs. Accumulation can be reduced through a more rigorous maintenance schedule.

13. Sludge accumulation in the WRS. Accumulation can be reduced through a more rigorous maintenance schedule.

14. Fouling in the WRS and BRS. Fouling can be reduced through a more rigorous maintenance schedule.

15. Filter cake has too much liquid. The filter cake can be shipped to a TSDF in a wet-state, but more shipments may be the result.

16. Bio-solids in BRS result in foaming and fouling. Bio-solids can be removed through a more rigorous maintenance schedule.

ACWA, Bechtel Pueblo Team and the CO CAC are of one mind to ensure that the costs to construct the BTA were dollars well spent and that the choice of neutralization followed by bio-treatment was a wise choice made by the community and the Department of Defense. While no one wants to minimize these issues, there is also a determination that all of the potential concerns listed above can and will be resolved to the satisfaction of all stakeholders based on the vast knowledge of those employed to make the BTA a successful part of this process.

Exhibit 4: Continued.

C

Biographical Sketches of Committee Members

ROBERT A. BEAUDET, *Co-Chair*, is an emeritus professor of chemistry. In 2005 he retired from the faculty of the University of Southern California, where he had served continuously in the Department of Chemistry since 1962. He has also been a part-time employee at Caltech's Jet Propulsion Laboratory since 1983. He received his Ph.D. in physical chemistry from Harvard University in 1962. From 1961 to 1962, he was a U.S. Army Chemical Corps officer and served most of his tour of duty at the Jet Propulsion Laboratory as a research scientist. Dr. Beaudet since has served on Department of Defense committees addressing both offensive and defensive considerations surrounding chemical warfare agents. He was chair of an Army science board committee that addressed chemical detection and trace gas analysis. He was also the chair of an Air Force technical conference on chemical warfare decontamination and protection. He has participated in two NRC studies on chemical and biological sensor technologies and energetic materials and technologies. Most of his academic career has been devoted to research in molecular structure and molecular spectroscopy. Previously, Dr. Beaudet served as a member of the NRC's Board on Army Science and Technology (BAST), as a member of the NRC Committee on Review of the Non-Stockpile Chemical Materiel Disposal Program, and as a BAST liaison to the Committee on Review and Evaluation of the Army Chemical Stockpile Disposal Program (Stockpile Committee). Since 1996, he has chaired or served as a member on various NRC committees reviewing the design evolution of the Assembled Chemical Weapons Alternatives program for the pilot plant facilities in Colorado and Kentucky.

TODD A. KIMMELL, *Co-Chair*, is principal investigator with the Environmental Science Division at the U.S. Department of Energy's Argonne National Laboratory. He is an environmental scientist and policy analyst, with more than 30 years' experience in solid and hazardous waste management, permitting and regulatory compliance, cleanup programs, environmental programs policy development, and emergency management and homeland security. He has supported the Army's chemical and conventional munitions management programs, and has contributed to the Army's Assembled Chemical Weapons Assessment Program and the Chemical Stockpile Emergency Preparedness Program. Mr. Kimmell also has a strong technical background in analytical and physical/chemical test method development and analytical quality assurance and control. Mr. Kimmell has also supported a number of environmental permitting programs at Army chemical weapons storage sites and at open burning/open detonation sites. He graduated from George Washington University with an M.S. in environmental science.

EDWARD BOUWER is currently the Abel Wolman Professor of Environmental Engineering and chair of the department of Geography and Environmental Engineering at Johns Hopkins University. He is also director of the Center for Contaminant Transport, Fate and Remediation. Before that, Dr. Bouwer spent 7 years as director of the Center for Hazardous Substances in Urban Environments, a project that was funded by the U.S. Environmental Protection Agency. Dr. Bouwer's research interests encompass factors that influence biotransformation of contaminants; bioremediation for control of contaminated soils and groundwaters; biofilm kinetics; biological processes design in wastewater, industrial, and drinking water treatment; transport and fate of microorganisms in porous media; and the behavior of metal and organic contaminants in sediments and aquatic ecosystems. Dr. Bouwer received a B.S.C.E. in civil engineering with a minor in nuclear engineering from Arizona State University and an M.S. and a Ph.D. in environmental engineering and science from Stanford University.

JUDITH A. BRADBURY graduated from the University of Pittsburgh with a Ph.D. in public and international affairs and has an M.A. in public affairs from the Indiana University of Pennsylvania and a B.S. in sociology from the London

School of Economics. She retired after almost 20 years as a senior social scientist with the Pacific Northwest National Laboratory, which is operated by Battelle for the U.S. Department of Energy. In her work, she has emphasized the relevance of social science insights and tools to the analysis and resolution of science policy issues. She has extensive experience in both the practice and research of public involvement and institutional activities. Her experience includes (most recently) responsibility for planning and implementing outreach and education activities for the Midwest Regional Carbon Sequestration partnership. Previous work includes evaluation of selected U.S. Army restoration advisory boards; a series of evaluations of the effectiveness of DOE's site-specific advisory boards; evaluation of training programs in public participation for DOE managers; meeting facilitation, planning, and program evaluation for the DOE nuclear waste transportation program; and research into community perspectives of the risk of incineration for disposing of the nation's stockpile of chemical weapons.

REBECCA A. HAFFENDEN, ESQ., is an attorney and currently serves as a program attorney at the Argonne National Laboratory. Her recent professional work has included work for the U.S. Department of Homeland Security to evaluate legislation and regulations associated with security vulnerabilities and providing legal expertise to programs involving federal facility site remediation and hazardous waste compliance and corrective actions (RCRA). She also coauthored a working paper on the application of federal and state hazardous waste regulatory programs to waste chemical agents, in addition to being a coauthor of the Environmental Impact Statement for the Assembled Chemical Weapons Alternates program. Ms. Haffenden received a B.A. in psychology from the University of Illinois and a J.D. degree from Suffolk Law School in Boston.

HANK C. JENKINS-SMITH is a professor of political science and associate director for the Center for Applied Social Research at the University of Oklahoma. Previously, he was a professor of public policy at the George Bush School of Government and Public Service at Texas A&M University. Before that, he served as an associate professor in the Department of Political Science at the University of New Mexico (UNM) and director of the UNM Institute for Public Policy. His areas of expertise include statistical analysis, public opinion measurement, politics of risk perception, environmental policy, and public policy. Dr. Jenkins-Smith is a member of the Society for Risk Analysis (SRA) and the American Political Science Association. In 1996, he received the SRA's Risk Research Award. He has over 60 publications and reports in his areas of expertise. Dr. Jenkins-Smith received his Ph.D. in political science from the University of Rochester.

KIMBERLY L. JONES is a professor and chair of the Department of Civil Engineering at Howard University. Before that, from 1996 to 2009, she worked as an associate and assistant professor in the department. Over the past 5 years, her research objectives have primarily been interdisciplinary, collaborative research in the emerging research areas of nanotechnology and nanobiotechnology, while continuing to build her environmental engineering capabilities. She has worked to develop an effective research strategy to investigate innovative technologies involving nanotechnology, environmental engineering, and membrane processes in an effort to solve some of the more pervasive problems facing our world, while working to attract, retain, and graduate technically competent African-American students to increase the number of minority engineers and scientists in academic, industrial, and government-related careers. Dr. Jones received her B.S. in civil engineering from Howard University, an M.S. in civil and environmental engineering from the University of Illinois, and a Ph.D. in environmental engineering from the Johns Hopkins University.

MURRAY GLENN LORD is associate environmental health and safety (EH&S) director in the EH&S Operations Technology Center at Dow Chemical Company. He is responsible for research programs for technology development for global environmental operations, which includes project areas in process optimization, technology development, and capital project execution. Mr. Lord has experience in project areas across multiple business and technology areas. He is also accountable for EH&S performance, budget performance, project development, and personnel leadership of research groups at four locations and is the leader of the Environmental Technology Leadership Group, accountable for environmental technology development for Dow. Previously, Mr. Lord was a technical leader of propylene oxide process research and was responsible for a research program in support of technology development for the propylene oxide process. He was also responsible for development and coordination of research studies at laboratory, pilot plant, and full commercial scale.

TRISHA H. MILLER is a systems analyst/engineer with Sandia National Laboratories. She has participated in a number of analysis projects focused on chemical security, including projects supporting the Department of Homeland Security related to evaluation of the security benefits of inherently safer technologies in the chemical industry and risk assessments for chemical attacks. Dr. Miller was awarded an early-career grant to develop new methodologies for end-to-end analysis of chemical defense systems. She received her Ph.D. in chemistry from the University of California at Berkeley in 2009. She serves as an adjunct faculty member in the Department of Chemistry at Augsburg College in Minneapolis.

ROBERT B. PUYEAR is an independent consultant specializing in corrosion prevention and control, failure analysis, and materials selection. Mr. Puyear worked at the Haynes Stellite Division of Union Carbide for 16 years developing high-performance materials for chemical and aerospace applications. He also worked for Monsanto for 21 years as a corrosion specialist, where he managed the Mechanical and Materials Engineering Section. He is an expert in materials engineering and evaluating materials of construction. Mr. Puyear graduated from the Missouri School of Mines and Metallurgy with a B.S. in chemical engineering and from Purdue University with an M.S. in industrial administration. He was also a member on the National Research Council Committee on Review and Evaluation of the Army Chemical Stockpile Disposal Program.

WILLIAM R. RHYNE is a retired risk and safety analysis consultant to the nuclear, chemical, and transportation industries. He has over 30 years' experience associated with nuclear and chemical processing facilities and with the transportation of hazardous materials. From 1984 to 1987, he was the project manager and principal investigator for a probabilistic analysis of transporting obsolete chemical munitions. Beginning in1997, he has been a member of several NRC committees for the Assembled Chemical Weapons Alternatives program and is a former member of the Committee on Chemical Demilitarization (2007-2010). Dr. Rhyne has authored or coauthored numerous publications and reports on nuclear and chemical safety and risk analysis areas and is the author of the book *Hazardous Materials Transportation Risk Analysis: Quantitative Approaches for Truck and Train.* He is a former member of the NRC Transportation Research Board's Hazardous Materials Committee, the Society for Risk Analysis, the American Nuclear Society, and the American Institute of Chemical Engineers. He received a B.S. in nuclear engineering from the University of Tennessee and M.S. and D.Sc. degrees in nuclear engineering from the University of Virginia.

PHILLIP E. SAVAGE is the head of the Chemical Engineering Department at Penn State. He earned a B.S. from Penn State and M.Ch.E. and Ph.D. degrees from the University of Delaware. All of his degrees are in chemical engineering. His research and teaching focus on the rates, the mechanisms, and the engineering of chemical reactions that move us toward a more environmentally sustainable society. Current research projects deal with hydrothermal reactions that can be used for hydrogen production from biomass and for liquid transportation fuel production from algae. His research group uses experiments, modeling, and simulation to explore different reaction systems. His teaching focuses on chemical reaction engineering and environmental sustainability. Dr. Savage is editor-in-chief for *Industrial & Engineering Chemistry Research,* and he is on the editorial boards for the *Journal of Supercritical Fluids, Energy & Fuels,* and *Environmental Progress & Sustainable Energy.* He has served as chair of the Catalysis & Reaction Engineering division of the American Institute of Chemical Engineers (AIChE) and the Industrial & Engineering Chemistry division of the American Chemical Society (ACS). Dr. Savage is a fellow of the AIChE and the ACS. He received the 2009 Michigan Governor's Award for Green Chemistry and the 2001 National Catalyst Award from the American Chemistry Council in recognition of his outstanding teaching and contributions to chemical education.

PHILIP C. SINGER is an emeritus professor in the Department of Environmental Sciences and Engineering in the Gillings School of Global Public Health at the University of North Carolina (UNC) at Chapel Hill, where he was the Dan Okun Distinguished Professor of Environmental Engineering from 2002 to 2010. After obtaining his Ph.D. from Harvard University in 1969, Dr. Singer was an assistant professor in the Department of Civil Engineering at the University of Notre Dame before joining the faculty at UNC in 1973. He conducted research on the chemical aspects of water and wastewater treatment and on aquatic chemistry for 45 years and has published more than 250 papers and reports on these subjects. Dr. Singer has been active in the American Water Works Association and has served on the National Research Council's Water Science and Technology Board and its Committee on Drinking Water Contaminants, as well as on the U.S. Environmental Protection Agency's Science Advisory Board and the National Drinking Water Advisory Council. He is a recipient of the American Water Works Association's A.P. Black Research Award and the Abel Wolman Award of Excellence, the American Academy of Environmental Engineers' Gordon Maskew Fair Award, the National Water Research Institute's Athalie Richardson Irvine Clarke Prize, and the Association of Environmental Engineering and Science Professor's Charles R. O'Melia Distinguished Educator Award. He was elected to membership in the National Academy of Engineering in 2005. Dr. Singer is currently a part-time consultant with CDM-Smith.

D

Committee Meetings

FIRST MEETING
JULY 29-31, 2014
CONVENTION CENTER
PUEBLO, COLORADO

Objective: To introduce required administrative procedures set forth by the National Research Council; conduct the composition and balance discussion; read the committee statement of task and background review with committee sponsor; receive briefing presentations on the demilitarization process at the Pueblo Chemical Agent Destruction Pilot Plant; review preliminary report outline and report-writing process; confirm committee writing assignments and discuss next steps and future meeting dates.

Safety Briefing and Site Tour of the Pueblo Chemical Agent Destruction Pilot Plant

Study Origin and Context, Bruce Huenefeld, PCAPP site project manager

PCAPP Project Overview, Bret Griebenow, PCAPP deputy project manager

PCAPP's Agent Treatment Process, Bill Steedman, PCAPP senior process engineer

Committee Approach to Soliciting Stakeholder Input, Judith Bradbury, Committee on Hydrolysate, member

Stakeholder Interests and Views on Offsite Shipment of Hydrolysate: Key Issues at the Site in Past, Miguel Monteverde, public affairs specialist, PEO ACWA, and Sandy Romero, PCAPP communications manager

Stakeholder Interests and Views on Offsite Shipment of Hydrolysate, Irene Kornelly, chair, Citizens' Advisory Commission

Biotreatment Process: Cradle to Grave, Paul Usinowicz, PCAPP technical advisor; Yakup Nurdogan, PCAPP lead industrial wastewater treatment engineer

Biotreatment Risk Mitigation Activities, Don Guzzetti, PCAPP start-up field supervisor, and Dave deLesdernier, Battelle support

RCRA Permit Structure and Potential Modifications for Offsite Shipment of Hydrolysate, Ron Entz, PCAPP environmental permitting manager

Other Regulatory Requirements and Notifications Applicable to Offsite Shipment of Hydrolysate, Paul Warbington, PCAPP environmental manager

PCD RCRA Permit from Perspective of the CDPHE Regulatory and Other Requirements and Notifications Applicable to Offsite Shipment, Doug Knappe, permitting unit leader for PCAPP, and Joe Schieffelin, Hazardous Waste Management Program manager, Colorado Department of Public Health and Environment

GB Bomblet Disposal at Rocky Mountain Arsenal: A Case History, Laurence Gottschalk, director, Recovered Chemical Materiel Directorate (remote presentation)

Internal Army Requirements That May Apply If Hydrolysate Is Sent Offsite, and Case Studies of Offsite Shipment of Hydrolysate, Amy Dean, environmental engineer, Office of the Program Manager for the Elimination of Chemical Weapons (remote presentation)

Downstream Impacts to Plant Operations If Offsite Shipment Is Required (Discussion), Facilitator: George Lecakes, PCAPP chief scientist

CITIZENS' ADVISORY COMMISSION PERMITTING WORKING GROUP MEETING JULY 30, 2014 DISTRICT ATTORNEY'S CONFERENCE ROOM PUEBLO, COLORADO

Becca Haffenden and Nia Johnson

CITIZENS' ADVISORY COMMISSION MONTHLY MEETING JULY 30, 2014 OLDE TOWNE CARRIAGE HOUSE PUEBLO, COLORADO

National Research Council, Committee on Hydrolysate

Committee Approach to Soliciting Stakeholder Input, Robert A. Beaudet, chair, and Judith Bradbury, member, Committee on Hydrolysate

Committee on Hydrolysate members

TELECONFERENCE WITH PCAPP PUBLIC AFFAIRS OFFICE SEPTEMBER 5, 2014

Participants in Teleconference with Public Affairs Personnel:

Judith Bradbury, member, Committee on Hydrolysate
Hank Jenkins-Smith, member, Committee on Hydrolysate
Irene Kornelly, Chair, Colorado Citizens' Advisory Commission
Jeannine Natterman, Colorado Department of Public Health and Environment
John Norton, member, Colorado Citizens' Advisory Commission
Sandy Romero, PCAPP communications manager
Nancy Schulte, study director, NRC
Thomas Schultz, public affairs specialist, PCAPP

SECOND MEETING SEPTEMBER 9-11, 2014 NATIONAL ACADEMY OF SCIENCES WASHINGTON, D.C.

Objective: Conduct committee deliberations, discuss report, and discuss adequacy of data gathering to date; formulate plan to complete any necessary data gathering; decide and agree on findings and recommendations, conduct report drafting to achieve a preconcurrence draft, and make any necessary work assignments.

THIRD MEETING NOVEMBER 5-6, 2014 KECK CENTER WASHINGTON, D.C.

Objective: Conduct committee deliberations, discuss report, and develop and review any additional text as may be deemed necessary to complete the report; reach concurrence on report findings and recommendations; describe NRC peer review process as performed for DEPS/BAST reports.